Functional Reverse Engineering of Strategic and Non-Strategic Machine Tools

Computers in Engineering Design and Manufacturing

Series Editor:

Wasim Ahmed Khan

GIK Institute of Engineering Sciences & Technology, Topi, Pakistan

Functional Reverse Engineering of Machine Tools

Edited by Wasim Ahmed Khan, Ghulam Abbas, Khalid Rahman, Ghulam Hussain, Cedric Aimal Edwin

Functional Reverse Engineering of Strategic and Non-Strategic Machine Tools

Edited by Wasim Ahmed Khan, Khalid Rahman, Ghulam Hussain, and Ghulam Abbas

For more information on this series, please visit: https://www.routledge.com/Computers-in-Engineering-Design-and-Manufacturing/book-series/CRCCOMENGDES

Functional Reverse Engineering of Strategic and Non-Strategic Machine Tools

Edited by

Wasim Ahmed Khan, Khalid Rahman, Ghulam Hussain, and Ghulam Abbas

CRC Press
Taylor & Francis Group
Boca Raton London New York

CRC Press is an imprint of the
Taylor & Francis Group, an informa business

First edition published 2021
by CRC Press
6000 Broken Sound Parkway NW, Suite 300, Boca Raton, FL 33487-2742

and by CRC Press
2 Park Square, Milton Park, Abingdon, Oxon, OX14 4RN

CRC Press is an imprint of Taylor & Francis Group, LLC

Library of Congress Cataloging-in-Publication Data
Names: Khan, Wasim A., editor.
Title: Functional reverse engineering of strategic and non-strategic machine tools / edited by Wasim Ahmed Khan, Khalid Rahman, Ghulam Hussain, and Ghulam Abbas.
Description: First edition. | Boca Raton : CRC Press, 2021. | Series: Computers in engineering design & manufacturing | Includes bibliographical references and index.
Identifiers: LCCN 2020057799 (print) | LCCN 2020057800 (ebook) | ISBN 9780367365806 (hbk) | ISBN 9780367808235 (ebk)
Subjects: LCSH: Machine design. | Reverse engineering. | Technology transfer—Pakistan.
Classification: LCC TJ230 .F826 2021 (print) | LCC TJ230 (ebook) | DDC 621.9/02—dc23
LC record available at https://lccn.loc.gov/2020057799
LC ebook record available at https://lccn.loc.gov/2020057800

ISBN: 978-0-367-36580-6 (hbk)
ISBN: 978-1-032-02156-0 (pbk)
ISBN: 978-0-367-80823-5 (ebk)

Typeset in Times
by codeMantra

Dedication

To our Families

Contents

SECTION 4 Modeling and Simulation

Foreword

The state of industrial activity, in Khyber Pakhtunkhwa Province, Pakistan, especially the discrete manufacturing is very weak. Building the capacity to develop strategic and nonstrategic machine tools is crucial to have a strong industrial base in sectors such as machining, agriculture, forestry, irrigation, textile, biomedical, energy, and power. The capability built intends to produce, operate, and maintain these machine tools in start-ups, small-to-medium enterprises, and beyond. The program has a strong component of building critical mass well-trained manpower on computer-controlled machines to have minimum dependency on other nations.

Leading the Way – like in many Science, Technology, and Innovation areas, Directorate of Science and Technology (DoST), Department of ST and IT, Government of Khyber Pakhtunkhwa in collaboration with Ghulam Ishaq Khan Institute (GIKI) have launched 1st of its kind program of the development of the indigenous capability of Functional Reverse Engineering. The critical mass of well-trained manpower will cater for the needs of local, provincial, national, and international markets.

The programs besides import substitution, saving precious foreign exchange aims at contributing toward strengthening the position of Pakistan in the "Next Eleven Countries."

Dr. Khalid Khan
Director General/ Director DoST

Preface

Mechanization is key to productivity of small-to-medium enterprises comprising strategic and nonstrategic machine tools. Development of strategic and nonstrategic machine tools using system engineering design, RAMS, and Ergonomics involves teams of novice to expert human resource trained over a long period of time or their capacity built based on national requirement over a short period of time. As the development of strategic and nonstrategic machine tools is combination of many disciplines of Sciences and Technology, a directed approach to these purpose-built teams is necessary.

Rightfully, the call for proposal for this book enlists these areas and seeks continuity from the previous book in the series "Functional Reverse Engineering of Machine Tools." At first sight, the topics of the chapters in two books seemingly look apart but the fact is that the thought process is converging toward pure machine tool development. So much so that a complete section comprising four chapters in this book is directed toward machine tool development while two more chapters in the book are directed toward the architecture of the machine tool. Rest of the chapters are based on sensors development, selection of sensors and actuators, and modeling and simulation. One complete section is devoted toward computational analysis of machine tools.

It is evident that the entrepreneurs in Pakistan have started taking interest in the development of machine tool controllers as there is little investment requirement for printed circuit board development and industrial software development. Craftsmanship needed for material development and mechanical artifacts development needs immediate attention and manpower must be trained locally and overseas for rapid growth in this area.

It is expected that this intervention, introduced by Directorate of Science and Technology (DoST), Government of Khyber Pakhtunkhwa, Pakistan, shall be able to strengthen the position of Pakistan in the Next Eleven Countries.

The next edition of the series will give special consideration to additive manufacturing processes.

Editors
November 2020
GIK Institute of Engineering Sciences and Technology
Pakistan

Acknowledgments

The editors and contributors are indebted to Directorate of Science and Technology (DoST), Department of Science and Technology and IT, Government of Khyber Pakhtunkhwa, Pakistan for their continuous support and funding. Technical support from the Institute of Mechanical Engineers (IMechE), UK and Institution of Electrical and Electronics Engineers (IEEE), USA needs recognition.

The editors owe special thanks to all our contributors from China, USA, Turkey, Brunei, and Pakistan who faced all the difficulties posed by COVID-19 and went through a double-blind peer review. The 2nd International Workshop on Functional Reverse Engineering of Machine Tools still has to be organized due to the delays caused by COVID-19.

Volunteer work done by IMechE GIK Student Chapter needs mention. Special thanks to Dr. Memoon Sajid and Mr. Hassan Malik for their contribution during final review and formatting of the manuscript.

Editors

Wasim Ahmed Khan obtained his first degree in Mechanical Engineering from NED University of Engineering and Technology, Karachi, Pakistan. He later obtained a PhD degree in Operations Research from the Department of Mechanical Engineering, University of Sheffield, England, UK. He is a life member of Pakistan Engineering Council. He is also a chartered engineer (CEng) of the Engineering Council, UK, and a fellow of the Institution of Mechanical Engineers, UK. Professor Khan was recently elected as the senior member of Institution of Electrical and Electronics Engineers, USA. He has diverse work experience including working with manufacturing industry, software development for local and overseas clients and teaching production engineering, business, and computer science students. professor Khan is currently working as a professor in the Faculty of Mechanical Engineering, GIK Institute of Engineering Sciences and Technology. He is a reviewer of several internationally reputed journals. He is also a senior advisor, the catalyst (The GIK Incubator), and coordinator, GIK Institute Professional Education Program. Professor Khan also the founder of Advance Manufacturing Processes Laboratory (AMPL), Faculty of Mechanical Engineering, GIK Institute.

Khalid Rahman is an associate professor and dean in the faculty of Mechanical Engineering at GIK Institute of Engineering Sciences and Technology, where he has been a faculty member since 2012. He obtained his BS (Mechanical Engineering) from the GIK Institute and MS and PhD degrees from Jeju National University, South Korea, in 2012. He also has seven years of industrial experience in Engineering design and manufacturing. His research interests include direct wire technology for electronic devices and sensor fabrication and applications.

Ghulam Hussain is currently working as professor in the Faculty of Mechanical Engineering, GIK Institute of Engineering Sciences and Technology, Pakistan. His research interests include advanced manufacturing processes, plasticity, energy modeling, and industry 4.0. He is the author of numerous publications and stands among pioneers and top 10 leading researchers of die-less incremental forming processes. Based on his scientific contributions, he is listed in "Who is Who," and has been ranked among top 10 National Productive Scientists. He is actively involved in doing research with renowned international universities. He has been also selected as a foreign expert on Manufacturing in China. Moreover, he is a reviewer and editorial board member of several international reputed journals.

Ghulam Abbas received a BS degree in computer science from University of Peshawar, Pakistan, in 2003, a MS degree in distributed systems, and a PhD degree in computer networks from the University of Liverpool, UK, in 2005 and 2010, respectively. From 2006 to 2010, he was a research associate at Liverpool Hope University, UK, where he was associated with the Intelligent and Distributed

Systems Laboratory. Since 2011, he has been with the Faculty of Computer Sciences and Engineering, GIK Institute of Engineering Sciences and Technology, Pakistan. He is currently working as associate professor and director of the Huawei Network Academy. Dr. Abbas is a co-founding member of the Telecommunications and Networking (TeleCoN) Research Lab at GIK Institute. He is a fellow of the Institute of Science and Technology, UK, a fellow of the British Computer Society, and a senior member of the IEEE. His research interests include computer systems architecture and wireless and mobile communications.

Contributors

Muhammad Aamir
School of Engineering
Edith Cowan University
Joondalup, Australia

A. E. Pg. Abas
Faculty of Integrated Technologies
Universiti Brunei Darussalam
Gadong, Brunei

M. Kamran Abbasi
Faculty of Mechanical Engineering
GIK Institute of Engineering Sciences
and Technology
Topi, Pakistan

Sarmad Afzal
Faculty of Mechanical Engineering
GIK Institute of Engineering Sciences
and Technology
Topi, Pakistan

Iftikhar Ahmad
Faculty of Mechanical Engineering
GIK Institute of Engineering Sciences
and Technology
Topi, Pakistan

Shajee Ahmed
Faculty of Mechanical Engineering
GIK Institute of Engineering Sciences
and Technology
Topi, Pakistan

Hamza Altaf
Faculty of Mechanical Engineering
GIK Institute of Engineering Sciences
and Technology
Topi, Pakistan

Salman Amin
Faculty of Mechanical Engineering
GIK Institute of Engineering Sciences
and Technology
Topi, Pakistan

Ella M. Atkins
Department of Aerospace Engineering
University of Michigan Ann Arbor
Ann Arbor, Michigan

Kaan Buyuktas
Department of Mechanical Engineering
Middle East Technical University
Ankara, Turkey

Quentin Cheok
Faculty of Integrated Technologies
Universiti Brunei Darussalam
Gadong, Brunei

Sarmad Chohan
Faculty of Mechanical Engineering
GIK Institute of Engineering Sciences
and Technology
Topi, Pakistan

Volkan Esat
Department of Mechanical Engineering
Middle East Technical University
Ankara, Turkey

Shehroze Faisal
Faculty of Mechanical Engineering
GIK Institute of Engineering Sciences
and Technology
Topi, Pakistan

Muhammad Faizan
Faculty of Engineering, Science, and Technology (FEST)
Hamdard University
Karachi, Pakistan

Syed Saad Farooq
Department of Mechanical Engineering
Khwaja Fareed University of Engineering and IT
Rahim Yar Khan, Pakistan

Muhammad Umer Farooq
Department of Mechanical Engineering
Khwaja Fareed University of Engineering and IT
Rahim Yar Khan, Pakistan

Abdullah Haroon
Faculty of Mechanical Engineering
GIK Institute of Engineering Sciences and Technology
Topi, Pakistan

Muhammad Hasan
Faculty of Mechanical Engineering
GIK Institute of Engineering Sciences and Technology
Topi, Pakistan

Ning He
College of Mechanical and Electrical Engineering
Nanjing University of Aeronautics and Astronautics
Nanjing, People's Republic of China

Asif Iqbal
Faculty of Integrated Technologies
Universiti Brunei Darussalam
Gadong, Brunei

Muhammad Iqbal
Faculty of Integrated Technologies
Universiti Brunei Darussalam
Gadong, Brunei

Sohaib Jabran
Faculty of Mechanical Engineering
GIK Institute of Engineering Sciences and Technology
Topi, Pakistan

Muhammad Jamil
College of Mechanical and Electrical Engineering
Nanjing University of Aeronautics and Astronautics
Nanjing, People's Republic of China

Mazhar Javed
Faculty of Electrical Engineering
GIK Institute of Engineering Sciences and Technology
Topi, Pakistan

Waqar Joyia
Department of Mechanical Engineering
Middle East Technical University
Ankara, Turkey

Kemran Karimov
Department of Mechanical Engineering
Middle East Technical University
Ankara, Turkey

Zareena Kausar
Department of Mechatronics Engineering
Air University
Islamabad, Pakistan

Aqib Mashood khan
Department of Mechanical Manufacture and Automation
Nanjing University of Aeronautics and Astronautics
Nanjing, People's Republic of China

Farid U. Khan
Institute of Mechatronics
University of Engineering and Technology
Peshawar, Pakistan

Ghias Mahmood Khan
Department of Mechanical Engineering
Khwaja Fareed University of Engineering and IT
Rahim Yar Khan, Pakistan

Ihtesham Khan
Department of Mechanical Engineering
Middle East Technical University
Ankara, Turkey

Mohammad Zainullah Khan
Faculty of Mechanical Engineering
GIK Institute of Engineering Sciences and Technology
Topi, Pakistan

Muhammad Faisal Khan
Faculty of Engineering, Science, and Technology (FEST)
Hamdard University
Karachi, Pakistan

Zhang Liyan
Department of Engineering
Nanjing University of Aeronautics and Astronautics
Nanjing, People's Republic of China

Ali Nasir
Department of Electrical Engineering
University of Central Punjab
Lahore, Pakistan

Malik M. Nauman
Faculty of Integrated Technologies
Universiti Brunei Darussalam
Gadong, Brunei

Kamran Nazeer
Department of Mechanical Engineering
Khwaja Fareed University of Engineering and IT
Rahim Yar Khan, Pakistan

K. Rehman
Faculty of Mechanical Engineering
GIK Institute of Engineering Sciences and Technology
Topi, Pakistan

Rafay Safdar
Faculty of Mechanical Engineering
GIK Institute of Engineering Sciences and Technology
Topi, Pakistan

Memoon Sajid
Faculty of Electrical Engineering
GIK Institute of Engineering Sciences and Technology
Topi, Pakistan

Adeem Samad
Faculty of Mechanical Engineering
GIK Institute of Engineering Sciences and Technology
Topi, Pakistan

Muhammad Faizan Shah
Department of Mechanical Engineering
Khwaja Fareed University of Engineering and IT
Rahim Yar Khan, Pakistan

Mohammad Shahrukh
Faculty of Mechanical Engineering
GIK Institute of Engineering Sciences and Technology
Topi, Pakistan

Muftooh Ur Rehman Siddiqi
Department of Mechanical Engineering
CECOS University of Emerging Science and Information
Peshawar, Pakistan

Muhammad Suhaib
Department of Mechanical Engineering
Middle East Technical University
Ankara, Turkey

M. Suleman
Department of Mechanical Engineering
CECOS University of Emerging Science and Information
Peshawar, Pakistan

Faizan Tariq
Faculty of Mechanical Engineering
GIK Institute of Engineering Sciences and Technology
Topi, Pakistan

Sundus Tariq
Department of Mechanical Engineering
CECOS University of Emerging Science and Information
Peshawar, Pakistan

Sami Ullah
Faculty of Mechanical Engineering
GIK Institute of Engineering Sciences and Technology
Topi, Pakistan

Li Xiaojun
AVIC Chengdu Aircraft Industrial (Group) Co. Ltd.
Chengdu, People's Republic of China

Lin Yutao
Department of Engineering
Nanjing University of Aeronautics and Astronautics
Nanjing, People's Republic of China

Fatima Mohsin Zakai
Faculty of Engineering, Science, and Technology (FEST)
Hamdard University
Karachi, Pakistan

Cheng Zhang
Department of structural engineering and mechanics
Nanjing University of Aeronautics and Astronautics
Nanjing, People's Republic of China

Wei Zhao
College of Mechanical and Electrical Engineering
Nanjing University of Aeronautics and Astronautics
Nanjing, People's Republic of China

Lv Zhengyang
Department of Electrical Engineering
Nanjing University of Aeronautics and Astronautics
Nanjing, People's Republic of China

Section 1

Functional Reverse Engineering Activities in Pakistan

A strategic machine tool is a machine tool that can produce anything within its scope as well as parts of nonstrategic machine tools. The current intervention initiated 3years ago is having a strong effect on academia and entrepreneurs to adapt it. A number of groups have emerged within Pakistan seeking knowledge in the area for different engineering sectors and entrepreneurs are more willing to invest in the area. Federal Government, Government of Khyber Pakhtunkhwa, and number of national and international donors, despite current COVID-19 situations, are working on this stimulus.

Ghulam Ishaq Khan Institute of Engineering Sciences and Technology, Pakistan, is in the forefront of this activity and work done at the institute under the funding of Government of Khyber Pakhtunkhwa is striving for capacity building in the province and the country.

The knowledge being gained through experimentation at Advance Manufacturing Processes Laboratory (AMPL), Faculty of Mechanical Engineering, GIK Institute, is incremental and is increasing with time. It is at par with any international laboratory working in the same area.

The knowledge gained in the AMPL is disseminated through undergraduate, graduate, and postgraduate courses, Lectures and Webinars to student societies, professional bodies, and trade associations. One of the major knowledge dissemination tools is the International Workshops on Functional Reverse Engineering of Machine Tools.

1 3D Scanner
An Application of Functional Reverse Engineering

Mohammad Zainullah Khan, Muhammad Hasan, Abdullah Haroon, Mohammad Shahrukh, and Wasim Ahmed Khan
GIK Institute of Engineering Sciences and Technology

CONTENTS

1.1 INTRODUCTION

1.1.1 Background and Motivation

When one thinks of 3D scanning, usually the first thing that comes to mind is the design extraction of real-life objects onto a computer for further use. This technology allows us to rescale objects and to perform trial-and-error in fitting together complex geometries before doing so physically or being able to perform further modifications. It is also particularly useful for reverse engineering and rapid prototyping. For example, a 3D scanner can be used to replace a broken part of a desk lamp with the help of a 3D printer. The process is very simple. The broken part is first scanned, then modified (if required), and printed.

Since 3D scanners have limited availability and are mostly high-priced, we decided to develop one which fulfills the needs of an individual or a small start-up while being cost-effective. The scanner works by collecting the geometric data at different points, combining the result, and producing a three-dimensional image for further use. For every data point collected, additional information such as its color and texture may also be recorded. The initial points collected can also contain data of the distance from the camera to points on the object or this may be set manually. This allows for a common reference frame to be set up.

1.2 LITERATURE REVIEW

1.2.1 Literature Review

The applications for 3D scanning are immense, traversing a few extents of protest scale.[1] In a general sense, 3D scanning looks to digitize this present reality to enable it to be interfaced with, by programming. The correct meaning of 3D scanning is tricky. Some have looked to catch 3D articles to produce 2D motion pictures from them, conceivably making 3D models all the while. Others have made restricted 3D scans from a little gathering of picture focuses, considering spots set on the person detected. Numerous carefully vivified films have been made utilizing this point-based approach.

This technology also has importance in the field of image processing. This consists of reverse engineering, quality control,[2] and digitally conserving historical artifacts. Scanning has additionally been utilized to survey the bearing limit of a piece of concrete, without annihilating or harming it, as is required by a few techniques. Educational[3]

[1] 3D Scanning 101.
[2] Example: automotive products.
[3] e.g. Dalhousie University.

and commercial institutions[4] are also taking advantage of this technology by forming libraries of scanned objects, some of which can be found online. The work presented in the article[5] is approximately in view of the idea made by Richard Garsthagen for a Raspberry Pi-based examining framework that catches the majority of the pictures required for making a 3D display all the while.

1.2.2 Noncontact Active 3D Scanning

According to an article,[6] the scanning types can be divided into two categories, contact and noncontact. The noncontact type can be further divided into active and passive. The scanner presented in this chapter focuses on the active-type configuration. These active (dynamic) scanners emanate some sort of radiation or light and identify its appearance keeping in mind the end goal to test a condition. Conceivable kinds of emanations utilized include light, ultrasound, or X-rays.

3D laser scanning or 3D laser scanners can, for the most part, be arranged into three principle classes; time of flight, stage move, and laser triangulation. These laser filtering methods are normally utilized autonomously; however, they can likewise be utilized as a part of mix to make a more adaptable examining framework. There are additionally various other laser filtering innovations that are half and halves and additional mixes of other 3D examining advances, for example, accordion periphery interferometry, or conoscopic holography.

In accordion fringe interferometry, laser beams cast fringe patterns on objects, a digital camera captures images, and the software constructs point clouds of data. A software package then analyzes the point cloud to extract information such as dimensions or variations in shape for subsequent operations. Conoscopic holography is a new method for recording holograms using incoherent light as described. The method is based on optical propagation through birefringent crystals. Optical methods for the reconstruction of such a hologram are also presented.

1.2.3 Different Approaches to Noncontact Active 3D Scanning

1.2.3.1 Time-Of-Flight

The time-of-flight 3D laser scanner is a dynamic scanner that utilizes laser light to test the subject. At the core of this sort of scanner is a period of flight laser rangefinder. The laser rangefinder finds the separation of a surface by timing the round-trip time of a beam of light. A laser is utilized to radiate a beam of light and the measure of time before the reflected light is seen by a locator is coordinated. Since the speed of light is known, the round-trip time decides the movement separation of the light, which is double the separation between the scanner and the surface. The exactness of a period-of-flight 3D laser scanner relies upon how accurately the time can be estimated.[7]

---

[4] e.g. thingiverse.com.

[5] *Development of a Large, Low-Cost, Instant 3D Scanner.*

[6] 3D laser Scanners: History, Applications and Future.

[7] approximately $t = 3.3$ ps is the time taken for light to travel 1 mm.

The laser rangefinder just recognizes the separation of one point toward its view. Therefore, the scanner examines its whole field of view one point at any given moment by altering the range discoverer's course of view to filter distinctive focuses. The view course of the laser rangefinder can be changed either by pivoting the range discoverer itself or by utilizing an arrangement of turning mirrors. The last technique is generally utilized because mirrors are considerably lighter and would thus be able to be pivoted substantially quicker and with more noteworthy precision.

1.2.3.2 Phase Shift

The principle on which the phase shift laser scanners work on is the comparison of the shift in phase of reflected from the laser with the sample standard phase which is also captured. This process is like the process of measuring the distance using Time-of-Flight detection; however, the distance measurement through phase shift has a higher accuracy.

In the method discussed earlier, the sent and received signals are checked for the shift in phase, and thus, the distance can be calculated. The range of the certain modulation used for the measurement of the shift is half of the modulation wavelength. If high-frequency modulation is used, the range we get is smaller; however, we get high-frequency modulation which increases the precision. The ambiguity can be dealt with by using two different modulation frequencies. By using the phase difference for the computation from both frequencies, an unambiguous and precise measurement can be achieved.

1.2.3.3 Triangulation

In the triangulation method, as seen in Figure 1.1, laser light is used in the modern 3D scanners to perform the task of probing the environment. The process of triangulation is basically the targeting of the lasers onto a surface of the object needed to be

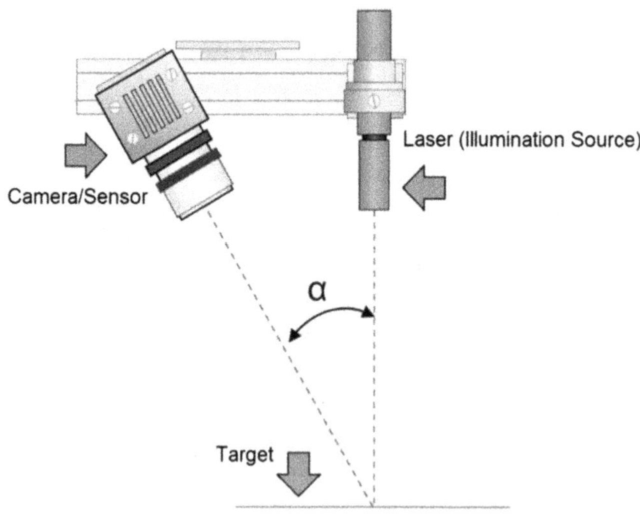

FIGURE 1.1 Triangulation method.

scanned. The camera reads the location pointed by the lasers and captures the image. The laser light that is reflected to the camera is from different points of the irregular surface of the object. A camera is used to estimate this distance. The camera, lasers, and the surface from which the laser bounces back form a triangle, hence the name, triangulation method. The distance between all these three is already known. The angle of the laser emitter is also known. With this known information, the shape and the size of the triangle can be recorded. Then by merging up all the points, we get the point cloud of the scanned object. This process was first experimented by The National Research Council of Canada in 1978.[8]

1.3 DESIGN AND ANALYSIS

1.3.1 Design Methodology

3D laser scanning is a technology that digitally captures the point data, with no physical contact being made. It is used to construct a similar digital replication of the physical object under observation. The physical dimensions of the digitally built 3D representation of the model are being the same as the physical object.

1.3.2 3D Scanning Process

1.3.2.1 Software

Open-source software "Free LSS" was used for image processing. It is written in C++ and licensed under the general public license (GPL). The scanning software runs self-contained on the Raspberry Pi without the need for a connected computer via USB; however, it is recommended to keep it connected to the computer either via LAN or Wi-Fi for greater flexibility in terms of settings. The software can be downloaded from http://www.freelss.org/.

1.3.2.2 Acquiring Data

First, the 3D scanner was powered on and configured by connecting it to the PC. This configuration process is required once per session and takes about 5 minutes. Once configured properly, the object that needs to be scanned is placed on the center of the turntable. A laser line was directed on the object, while a camera was used to continuously collect the data of alternating lasers and the shape of the laser lines as it sweeps along the object enabling the capturing of object contours to develop the 3D model.

1.3.2.3 Form of the Data

Data are acquired in the form of point cloud. A point cloud is a set of data points in three-dimensional coordinate system. The coordinates are often used to describe the external surfaces of the object and by convention are termed as *X*, *Y*, and *Z* coordinates.

---

[8] Roy Mayer (1999).

1.3.2.4 Further Processing of Data

Point cloud data attained were then automatically further processed to turn it into useable form. After this, an STL file was obtained which was moved to editing software for meshing, trimming, and formation of base.

1.3.2.5 CAD Model

At this point the 3D model of the scanned object was pretty much completed. Now, it could be exported to a CAD modelling software, such as Creo Parametric, Solid Works, etc., for further processing. Using these software packages, we can modify the 3D models and fix the geometry (if needed).

1.3.3 Governing Equations and Mathematical Modeling

1.3.3.1 Motor

The motor (located under the rotating platform) had to be capable of achieving 360-degree rotations with high precision of control and speed. Also, the torque at lower speeds had to be substantial. Using these two constraints, the best available option was to use a stepper motor instead of a servo DC motor.

To select the rating of the stepper motor, the maximum allowable weight of the scanable object was kept as the deciding parameter (10 kg) in this work. Also, the diameter of the turntable was chosen to be 26 cm. This size was chosen keeping in mind the moderately sized objects that would need to be scanned on an everyday basis. The calculations that were performed are as follows:

Diameter of turntable = 26 cm
Maximum weight of object to be scanned = 10 kg
Weight of turntable (approx.) = 7 kg

$$\text{Moment of Inertia of object } (I) = \frac{1}{2}MR^2$$

$$= \frac{1}{2} \times (10\,\text{kg}) \times (0.08)^2 \quad (1.1)$$

$$= 0.032\ \text{kg} \cdot \text{m}^2$$

$$\text{Moment of Inertia of turntable} = \frac{1}{2}MR^2$$

$$= \frac{1}{2} \times (7\,\text{kg}) \times (0.13)^2$$

$$= 0.05915\,\text{kg} \cdot \text{m}^2$$

$$\text{Total Moment of inertia} = 0.032\,\text{kg} \cdot \text{m}^2 + 0.05915\ \text{kg} \cdot \text{m}^2$$

$$= 0.09115\ \text{kg} \cdot \text{m}^2$$

$$\text{Tangential speed required} = 0.4 \text{ mm/s}$$

$$\text{Angular speed } (\omega) = \frac{0.4 \times 10^{-3}}{0.08} = 5 \times 10^{-3} \text{ rad/s}$$

$$\text{Time to accelerate} = 0.5 \text{seconds}$$

Required Torque:

$$\tau = \text{mass moment of inertia}(I)\text{angular acceleration}(\alpha) \tag{1.2}$$

$$\text{Angular acceleration } (\alpha) = \frac{\omega}{t} \tag{1.3}$$

$$\text{Torque}(\tau) = I \times \frac{\omega}{t}$$

$$= 0.01056 \text{ kg} \cdot \text{m}^2 \times \frac{5 \times 10^{-3}}{0.5} \tag{1.4}$$

$$= 0.1056 \text{ N} \cdot \text{mm}$$

To fulfill the torque and step angle accuracy (±5%) requirement, NEMA-17 stepper motor was selected. It is a bipolar stepper motor with 4-wire configuration; has 200 steps/revolution which are 1.8° per step. It has a maximum current of 350 mA and therefore it requires a stepper motor driver which is driven with two full H-bridges (Figure 1.2).

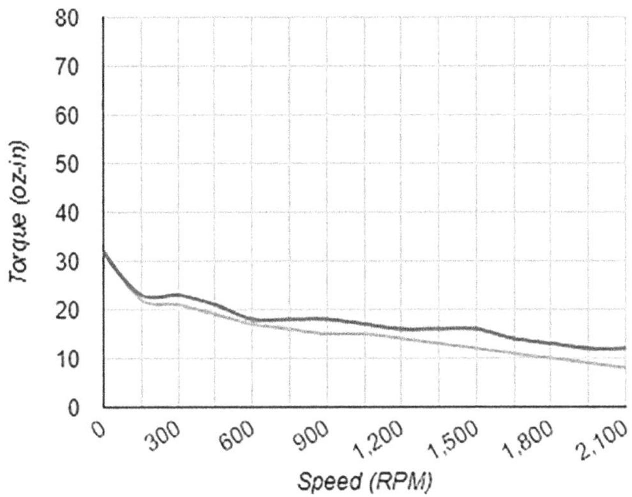

FIGURE 1.2 Torque vs RPM of stepper motor.

1.3.3.2 Laser

Two red LED line lasers were selected to be the transmitters. Their light was projected onto the target sample and the data points were captured by the camera. Camera and lasers were maintained at a fixed distance. First, a picture of the object was taken without any lasers being turn on. Then, one of the two lasers was switched on and another picture was taken. By computing the difference between the two images using a software, a laser trace was obtained. Red laser diodes were chosen since the high contrast makes it easier for the camera to pick up the contours of the object to be scanned. These lasers were also chosen to make scanning easier in daylight, meaning it would be easier for the camera to capture contours of the object on the turntable. Red LED also consumes the least amount of power in comparison with other LEDs of different colors.

Realizing that some of the irregular shapes are difficult to scan with a single laser, for example, a wine glass due to its concave part, it was decided to make use of two lasers. This helped to improve the quality because it reduces the number of "gaps" in the model due to the concave areas on the part.

One key area of concern is that lasers can cause eye damage if directed toward a person. Hence, one must be careful while testing and using the 3D scanner.

Model: Red Laser Diode 5V is easily available on amazon. They were purchased from a local online store in Pakistan (digipak online store).

1.3.3.3 Camera

The raspberry pi camera module selected for the 3D scanner was version 2. The camera was selected for its ability to produce a good-quality image with a high resolution of pixels. Furthermore, it gives the option of reducing the number of pixels so that the scanning time can be reduced (if desired).

The team's objective was to keep in mind that one of the main goals was to have the scanner as cost-effective as possible while fulfilling its purpose—hence, high-end expensive camera modules were not opted for. The specifications of the camera resolution are as follows.

$$\text{Active Pixel Count: } 3280\ (\text{Horizontal}) \times 2464\ (\text{Vertical})$$

1.3.3.4 Power Supply

The raspberry pi requires high performance to run the "Free LSS" software with the postprocessing required, hence needing a lot of power. As a result, building a all portable 3D scanner was not possible. The scanner was powered by a 12 V and 5 V AC to DC power adapter. One powered the 5 V raspberry pi and the other powered the 12 V stepper motor.

$$\begin{aligned}
&\text{The output current: } I = 2\,\text{A} \\
&\text{Power Produced from supply: } P = V \times I \\
&\text{Power (stepper motor): } 12\,\text{V} \times 2\,\text{A} = 24\,\text{W} \\
&\text{Power (RP3): } 5\,\text{V} \times 2\,\text{A} = 10\,\text{W} \\
&\text{Total Power: } 24 + 10\,\text{W} = 34\,\text{W}.
\end{aligned} \tag{1.5}$$

The raspberry pi takes 5 V and 1.8 A as input source, 50 mA to drive the general purpose input/output (GPIO) pins, and roughly 250 mA to drive the keyboard and mouse. As a result, the power consumption is

$$1.8\,\text{A} \times 5\,\text{V (raspberry pi2)} + 350\,\text{mA} \times 12\,\text{V (table)} + 2\,\text{A} \times 5\,\text{V (Camera)} = 23.2\,\text{W}.$$

The two power supplies selected were able to handle the power consumption.

1.3.4 Geometric Modelling and Design

The above image shows the complete assembly of the 3D scanner. The CAD model was designed using modelling software, Creo Parametric and Solid Works (Figure 1.3).

1.3.5 Analysis

Stress analysis was performed on both the turntable and the frame of the 3D scanner to ensure that they do not deform under the weight of the to-be-scanned object. The material selected for fabrication was ASTM A316 Steel as it is both economical and robust (factor of safety of 3) (Figures 1.4 and 1.5).

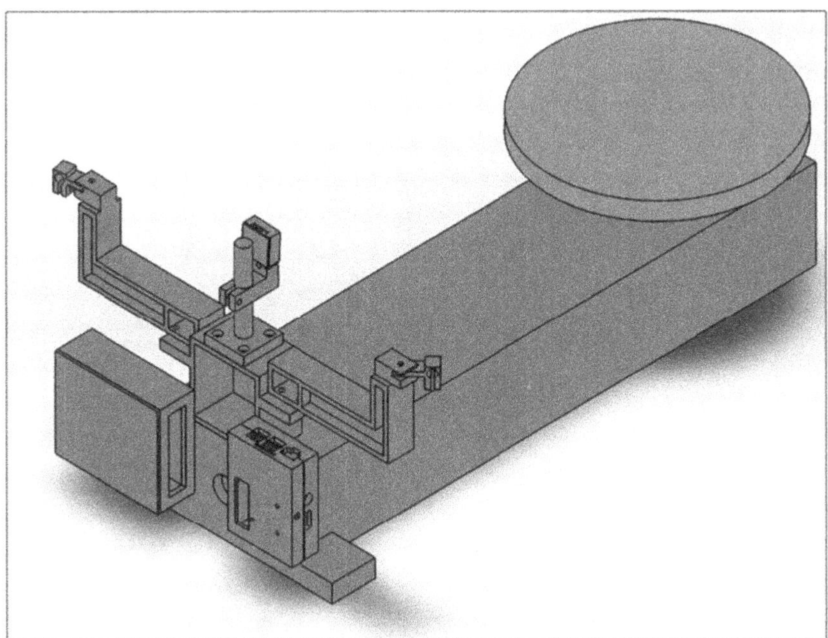

FIGURE 1.3 Finalized model.

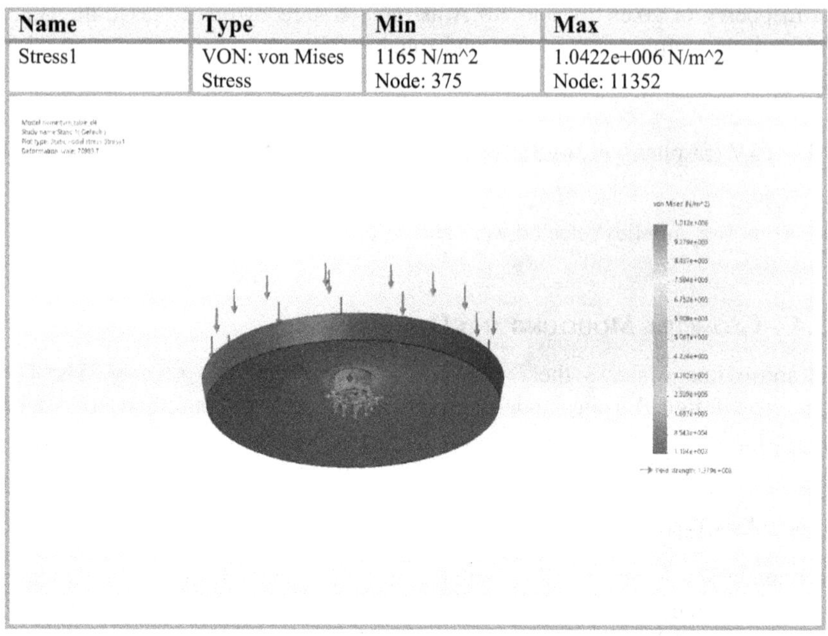

Name	Type	Min	Max
Stress1	VON: von Mises Stress	1165 N/m^2 Node: 375	1.0422e+006 N/m^2 Node: 11352

FIGURE 1.4 Stress analysis on turntable.

Name	Type	Min	Max
Stress1	VON: von Mises Stress	0.439682 N/m^2 Node: 2618	624905 N/m^2 Node: 15989

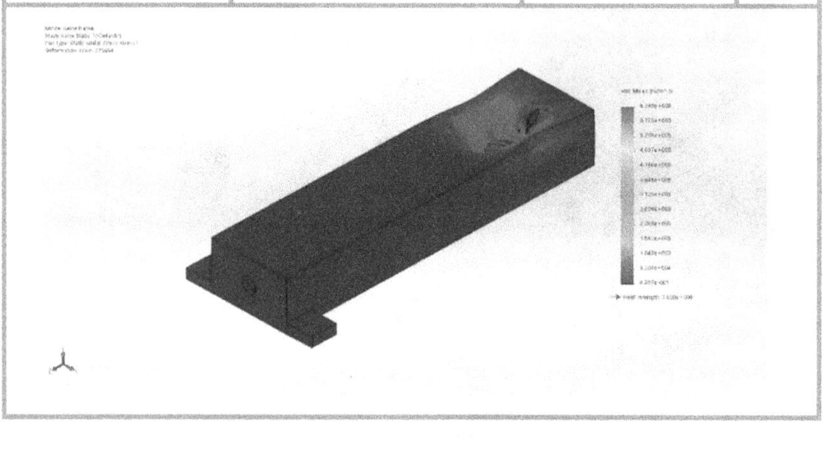

FIGURE 1.5 Stress analysis on frame.

1.4 IMPACT AND ECONOMIC ANALYSIS

1.4.1 Social Impact

Social impact refers not only to the degree to which the local community is affected but also considers the working class and the key stakeholders.

The consumers can take advantage of the applications of the 3D scanner in a wide variety of fields, namely, reverse engineering, industrial design and manufacturing, healthcare, education, and finally art and design. No matter what the application the 3D scanning is used for, at the end of the day, the consumer would have saved time and produced a better replica than using the old-manual time-consuming method of measuring each dimension and creating a CAD model.

3D scanning can transform and advance any business in the engineering, design and manufacturing, development, or testing domain. The technology is increasingly being adopted across the world, not only improving the speed of the average process cycle but greatly enhancing the overall quality of the product too. Put simply, 3D scanning captures digital data about the shape of a particular object using machinery and a light or laser to measure the distance between the scanner and the object itself with unerring accuracy. It is now being used to prototype complex parts of all shapes and sizes, from the tiniest cogs to a jumbo jet Airbus. One of the biggest benefits of 3D light scanning technology is the ability for businesses to work together efficiently to create multiple parts of a product. 3D model scanning makes it easier than ever to accurately measure and accommodate different parts into a finished product or prototype.

1.4.2 Sustainability Analysis

The use of this 3D scanner provides simple cost-effective scanning. Table 1.1 lists some of the 3D scanners available in the market today to give an idea of their cost and capabilities. This design also eliminates the need for any add-ons. The scans can be completed in 15–20 minutes, thus saving time and resources. Compared to the other commercially available scanners, this design has the advantage of being portable, low-cost, and easily upgradable (in terms of future software upgrades). Also, it has the capability to be controlled remotely if need be. The materials used in the fabrication of the 3D scanner mainly include ASTM 316 and Cast iron. Both materials are widely available and cost-effective. It was due to these two factors that these materials were selected after intense market research. Also, these materials can be easily substituted for Aluminum and ASTM 312. The ease of material substitution without any adverse effect on the performance makes this device easy to manufacture.

1.4.3 Environmental Impact

The materials that were selected for our design through analysis are ASTM 316 and cast iron which are completely recyclable materials and the manufacturing process for ASTM 316 parts is CNC milling which does not produce any harmful gases or harmful waste, all the waste in the form of iron chips can be recycled and the cooling agent used to cool the tool is also reused, hence reducing waste. Cast iron on the other

TABLE 1.1
Scanner Costs and Capabilities

Scanner	Cost	Type	Scanning Time
3D Systems Sense	$300	Optical—subject must move/ be moved	Varies by object/ operator
Artec EVA	$17,999	Optical—subject must move/ be moved	Varies by object/ operator
Geoma3ic Capture	$14,900	Optical with LED point emitter—subject must move/ be moved	Varies by object/ operator
Gotcha 3D Scanner	EUR 995	Optical—subject must move/ be moved	Varies by object/ operator
Head and Face Color 3D Scanner (Model 3030/RGB/PS) (CyEdit+)	$63,200	Laser—fixed scanner/fixed subject (supports the size of the front of a human head)	Not provided

hand produces slag, emissions (gasses and particulates), water pollution, and waste generation. Although established industries and manufacturing firms that follow standards to treat the waste before safely dumping or releasing to the atmosphere. It was made sure that only those manufacturing sites were contacted for hardware parts who followed environment-friendly processes.

1.4.4 Hazard Identification and Safety Measures

The model contains small parts and hence should be kept away from kids aged 5 and below. The microcontroller might tend to overheat while in continuous operation. If so, reset the device after a break of 15 minutes. The lasers are high intensity and should not be pointed directly toward the eye.

As far as the physical model is concerned, the maximum allowable weight is 10 kg. Any additional load might cause permanent damage and improper scans. The motor speed should not be manually fiddled with as increasing speed might cause incomplete images or object falling off from the turntable during the scanning phase.

1.5 LIMITATIONS

- It is difficult to scan glass, very shiny, or translucent objects as the laser is not properly reflected back to the camera. Powder-coating these might help but we were unable to test this out.
- The scanner finds it difficult to scan moving and nonsolid objects.

- Highly detailed objects require a lot of time for scanning and processing. This time can exceed a couple of hours, and even then, some of the details might be missed out.
- Sharp objects or parts are very difficult to scan such as a knife.
- The scans are black and white due to the limitation imposed by the camera.
- We could not find a proper gear that could fit with the motor to increase the accuracy of scanning by reducing the speed even further.
- Patented designs may not be reproduced before prior written approval from the patent owner.

1.6 FUTURE RECOMMENDATIONS

Although the design was made as optimal as possible and was redesigned multiple times after being discussed with industry professionals, there are still some modifications and updates that can be made later to improve the design and make it more accurate.

Starting with building an enclosure around the whole system to limit ambient light to as minimal as possible. Since the ambient light reduces the brightness of lasers and the camera detects based on the contrast of the lasers, so it is best to have bright lasers and little influence of ambient light. Adding to this a brighter laser can also be used.

We had selected ASTM 316 for our hardware parts that are manufactured, this material greatly increases the weight of the 3D Scanner, to reduce this weight we can select Aluminum alloy as an alternative and reduce the weight making it further portable. However, this would increase the cost.

To improve the imaging of the scanned object we can change the camera and use a better camera with more pixels than the current 8 mega pixels. Using a better camera would decrease the speed and take more time, to avoid this we can use a more powerful Pi alternative microcontroller.

We are currently giving the option of two CAD formats in our user interface: STL and PLY. In the future more options for the output formats such as IGES can be provided.

BIBLIOGRAPHY

Blais, F., Picard, M., & Godin, G. (2004). Accurate 3D acquisition of freely moving objects. *2nd International Symposium on 3D Data Processing, Visualisation, and Transmission*, 2004. 3DPVT 2004. 2004 Sep 9 (pp. 422–429). IEEE.

Goel, S., & Lohani, B. (2010, May 30). A motion correction technique for laser scanning of moving objects. *IEEE Geoscience and Remote Sensing Letters*. 11(1):225–8.

Guttentag, D. A. (2010, October). Virtual reality: Applications and implications for tourism. *Tourism Management*. 31(5):637–51.

Liu, K., Wang, Y., Lau, D. L., Hao, Q., & Hassebrook, L. G. (2010, March). Dual-frequency pattern scheme for high-speed 3-D shape measurement. *Optics Express*. 18(5):5229–44.

Murobo. (2016, January 27). *ATLAS 3D Scanner by Murobo*. Retrieved from thingiverse: https://www.thingiverse.com/thing:1280901

Murobo LLC. (2013). *ATLAS 3D - The 3D Scanner You Print and Build Yourself!* Retrieved from kickstarter: https://www.kickstarter.com/projects/1545315380/atlas-3D-the-3D-scanner-you-print-and-build-yourse

Murphy, L. (2012). *Case Study: Old Mine Workings. Subsurface Laser Scanning Case Studies*, 2(1), pp. 41–46.

Teutsch, C. (2007). *Model-based Analysis and Evaluation of Point Sets from Optical 3D Laser Scanners*. Shaker Verlag GmbH, Germany.

The Free 3D Printable Laser Scanning System. (2012). Retrieved from Freelss: http://www.freelss.org/

Tom Spendlove. (2015, January 16). *ATLAS 3D - Print and Build-It-Yourself 3D Scanner*. Retrieved from engineering: https://www.engineering.com/DesignerEdge/DesignerEdgeArticles/ArticleID/9393/ATLAS-3D--print-and-build-it-yourself-3d-scanner.aspx

Zhang, Y. W. (2011). *Superfast Multifrequency Phase-Shifting Technique with Optimal Pulse Width Modulation*. Retrieved from osapublishing: https://www.osapublishing.org/oe/abstract.cfm?uri=oe-19-6-5149

2 Functional Reverse Engineering of Universal Testing Machine with Enhanced Positioning and Reduced Power Consumption

Ghulam Hussain, Salman Amin, Wasim Ahmed Khan, Sami Ullah, K. Rehman, Rafay Safdar, Ghulam Abbas, and Hamza Altaf
GIK Institute of Engineering Sciences and Technology

CONTENTS

2.1 INTRODUCTION

Universal Testing Machine (UTM) belongs to category of machines that are used for material characterization. This machine can perform a variety of tests like tensile test, shear test, torsion test, compression test, and creep test, etc., depending upon its design and capability. These machines are used in almost every design industry to obtain characteristic properties of the material being used to manufacture products. UTMs tests provide quality control staff and engineers with valuable data that are used not only to characterize materials but can also be used for pass/fail testing of finished parts and assemblies [1]. UTMs have been in existence since 1800s in various forms. One of the primitive uses was to test the strength of steel to be used in steam power boilers. While many basic models of these machines were being developed from 1850 to 1880, the design was not standardized and a large number of the manufacturers were striving to build their own machine [2]. In 1880 a Philadelphia engineer Tinius Olsen, a Norwegian immigrant, designed and patented a machine that later became famous as the "Little Giant." Back then all the material tests required separate machine and only one test could be performed on one machine. Olsen's instrument was the first machine that could perform tensile, transverse, and compression tests all alone. Olsen's mechanism became the ancestor of all subsequent testing machines that started being produced worldwide, and on Feb 6, 1880, Olsen submitted an application for a patent "new and useful improvement in testing machines" and patent no. 228214 was given to him on June 1, 1880 [3]. Since then a lot of advancements have been made to the design of UTMs to improve the accuracy of results, to reduce cost, and improve control and software systems.

UTMs are not readily available in the market and cost a lot to import, so it was the need of the hour that a machine similar in design to conventional machines but with enhanced features must be developed locally to overcome this problem. Functional reverse engineering approach was used in this work. This chapter will highlight the methods and means to contribute toward the gaps related to this work. Reverse engineering is the process of utilizing man-made object and deconstructing it to reveal its design and architecture. This process also involves measuring an object and then reconstructing it as a 3D model and using different manufacturing techniques to have a physical form of it.

This work is focused on functional reverse engineering approach and improvements in the design and control of machine effectively while keeping the economic and environmental aspects in view. The task was very challenging as there was very little relevant research work on this machine. The machine was to be designed from the conceptual design phase and then moving up with its design analysis and then manufacturing. One of the major highlights of this work was the inclusion of additional improvements and features in the design of this machine to achieve the end objectives. The improvements in both mechanical and electrical designs of UTM were proposed and a comparison is presented for original design and modified design. In the mechanical design the belt and pulley mechanism, for the transmission of torque from motor pinion to ball screws, was replaced by gear assembly. This would remove the error in position measurement arising due to slippage of belt on pulleys. Moreover, in electronics area modification pertains to the type of actuating motor. Stepper motor was used instead of previously used servo motor, for better accuracy in position measurement. Furthermore, the power rating of the motor used in the designed machine was compared with the power rating of an available model in the market.

Analysis of carbon emissions in the environmental analysis shows that designed machine was also efficient in reducing carbon emissions to the environment. The methodology adopted to design this machine has also been discussed later in the chapter and the details of mechanical design and electronic circuit have been documented. At the end, analysis of carbon emission during manufacturing and operation of 50 kN machine has been performed. Once manufactured this machine can be used in local industries and academic institutes and save the hefty extra cost, they have to bear in the form of import duties on these types of machines.

Major components of UTM including mechanical structure, ball screws, load cell, extensometer, motor, motor driver, and other electronic circuitry components have been shown in the labelled diagram of complete machine in Figure 2.2. Testing in UTMs is done while adopting some standard procedures, i.e., size and shape of specimen should be in accordance to some standard, rate of loading/strain rate should be predefined, and frequency of data plotting should also be in line with the required accuracy of results. ASTM-E8 standard [4] was chosen for testing on designed UTM because it is acceptable in large number of countries around the world. Moreover, the designed machine controlled the rate of load application on test specimen through the speed of crosshead whose range was 1–100 mm/min. The speed of crosshead, that the user inputs, was adjusted by making stepper motor rotate at the required rpm.

2.2 METHODOLOGY

The approach taken to complete this project was to study the available UTMs and then use reverse engineering along with creativity to devise a pragmatic working solution. Conventionally, servo motors are used to power these machines but the stepper motor used in the new design had a lower electrical power rating which not only saves product and electricity costs but also makes the machine more environment friendly.

The overall design and the function of every individual component of UTM was carefully understood. After that the 3D-model of each of component of machine was designed using SolidWorks (2017) software. At first, exact dimensions of components were not known, so observation of the machine which was already available in the market gave an estimate of dimensions of machine components according to different load capacities.

As the design process is iterative, changes were made to the design intermittently to meet the conflicting requirements such as high safety and local manufacturing capability, compatibility of load cells and grips, and their availability and other make or buy decisions at the conceptual design stage. Ball screws were selected to be able to meet load-bearing requirements from supplier's catalog. Moreover, the decision to use stepper motor instead of servo motor was also made due to its high position accuracy and low power consumption in comparison to the servo motor which use encoders for positioning. Furthermore, the decision of using gear train instead of belt and pulley for the transmission was made considering the backlash errors associated with belt and pulley to avoid position measuring error.

Design of this machine was H-type; vertical tabletop or floor placing. This design was selected using design selection criteria. The methods used for concept design selection were Pugh Method and Measurement Scale Comparison. The selection parameters were as follows:

- Load Capacity
- Test Types
- Moveable Length
- Column Type

- Maintenance
- Reliability
- Overall Cost of machine

A prototype was also developed to test the working of designed electronic circuit of machine. Figure 2.1 shows the flow of signals or commands in the electrical circuit of designed machine.

2.3 ELECTRONIC CIRCUITRY

The components that we used for integration and interfacing of control systems of our machine are as follows and have been shown in Figure 2.2:

1. Microcontroller-Arduino
2. Nema 34-stepper Motor
3. Stepper motor driver
4. AC to DC power supply
5. Load cell
6. HX711 amplifier
7. Current sensors ACS712

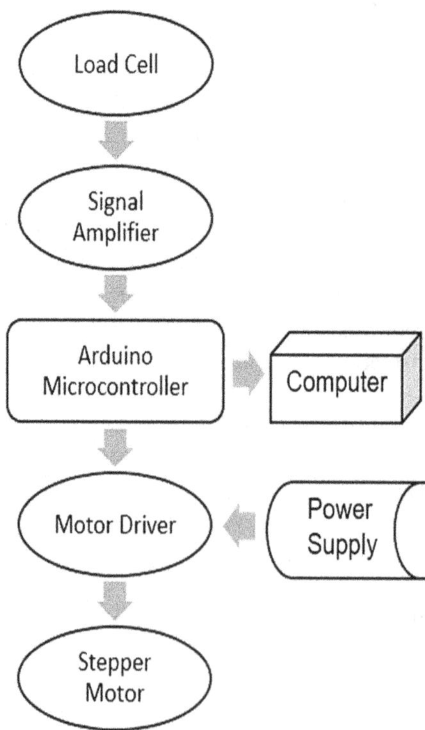

FIGURE 2.1 Electrical signal flow in UTM.

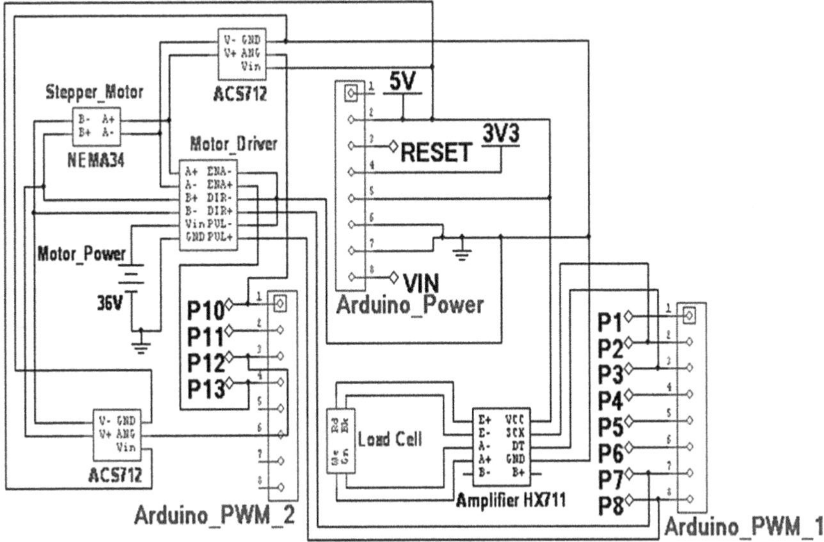

FIGURE 2.2 Electronic circuitry of designed machine.

2.3.1 Description of Connections

Arduino Mega was used for controlling motor motion and configuring load cell. NEMA 34 stepper motor has 4 wires that were connected to the motor driver through two ACS712 current sensor. ACS712 sensors also get 5 V power supply from Arduino. The load cell was connected to HX 711 amplifier. HX 711 amplifier was used to amplify signals of load cell as the signals produced by load cell are too weak to be detected by Arduino directly. HX 711 is connected to Arduino through SPI interface.

To control motion of motor, a code was written that converted the values for cross-head speed (entered by user) into the required rpm of motor shaft. Output of load cell was recorded through HX711 and Arduino and it can be plotted against extensometer data to have a stress–strain curve.

2.4 MECHANICAL DESIGN AND ANALYSIS

Machine structure with specimen gripped to its base was used in designing of a machine. The following design considerations were considered:

- Loading capacity: 50 kN
- Low cost
- Accuracy of results
- Number of tests to perform
- Conformance to standards
- Ease of maintenance

- Less friction and high wear resistance
- Less vibrations during operation
- Ease of relocation
- Ease of manufacture

Crosshead applies load to specimen by its upward or downward motion. Crosshead is basically a solid block of metal with housing space for load cell and flange nuts of ball screws. The motion of crosshead was speed controlled, and this was driven by the stepper motor for high accuracy in movements. The linear vertical motion of crosshead was provided by using ball screws, which convert rotational motion to linear motion. Guide supports were used for constraint motion of crosshead in vertical direction and to provide support to upper support plate, linear bearings, or bushings were used to prevent any unwanted friction during machine operation. Deep grove ball bearings were used to align ball screws vertically which were housed in the upper support plate. Model of machine with components labeled is shown in Figure 2.3.

A gear transmission mechanism was used to transfer rotational motion of motor to ball screws. It consists of triangular arrangement with one pinion (with smallest diameter), two idler gears, and two (with largest diameter) driven gears which are attached to ball screws. Assembly of gear transmission is shown in Figure 2.4.

The base of the machine was a solid block of metal with cavities for bearing housing for ball screw and guide supports. It was fixed to the frame placed on either floor or tabletop. The base design is shown in Figure 2.5.

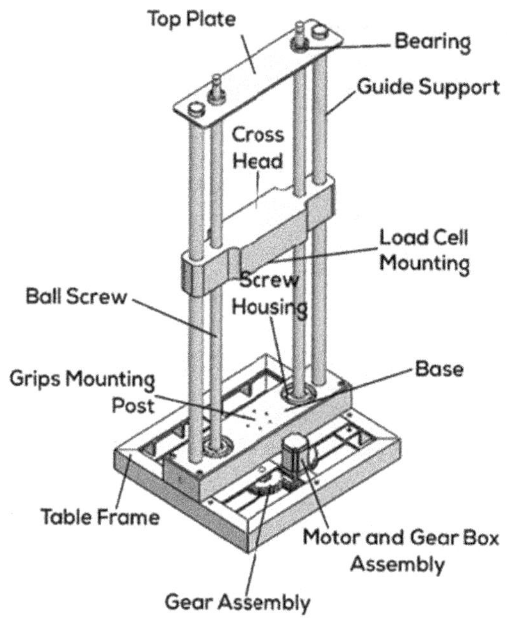

FIGURE 2.3 Labeled diagram of complete machine.

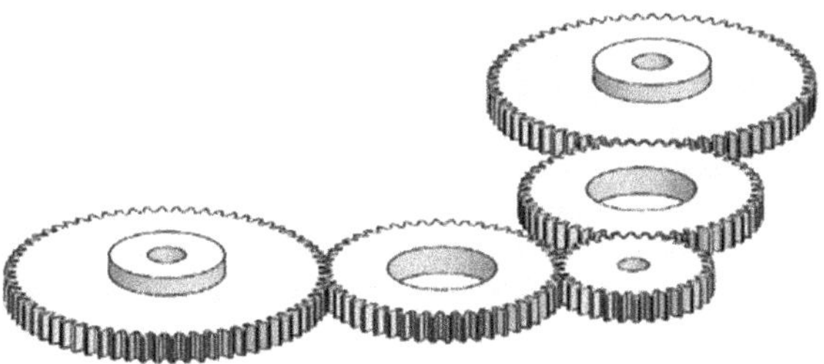

FIGURE 2.4 Gear train.

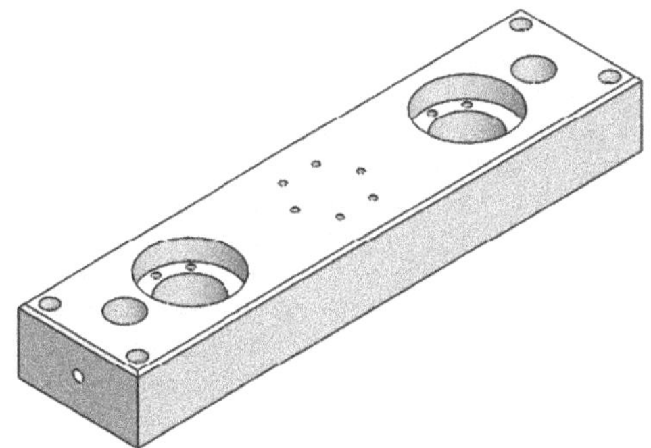

FIGURE 2.5 Base plate.

The major loads that arise during testing of materials on the crosshead and base were up to 50,000 N. To take this loading into account, static load analysis was performed using ANSYS. Load of 60,000 N was applied to determine factor of safety and bending stresses using Von-Mises failure criteria. The crosshead and base material used in analysis were AISI 1045 CD. The results are shown in Figures 2.6 and 2.7.

2.5 TESTING

A prototype of the machine was developed to test electronic systems and software. This prototype resembles the real design of machine in every aspect which was necessary for testing purpose. Specifications of the prototype are presented in Table 2.1.

FIGURE 2.6 FEA of base plate.

FIGURE 2.7 FEA of crosshead.

TABLE 2.1
Technical Specifications of the Developed Prototype

Characteristic	Description/Specification
Scaling	1:0.5
Material	Wooden Composite – MDF (Machine Frame) Mild Steel (Screws, Pulley, Motor Mount)
Load capacity	200 N
Actuating mechanism	Electromechanical (Belt and Pulley)
Orientation	Vertical
Software	MATLAB GUI
Micro controller	ARDUINO Mega
Specimen	Lead Wire, Copper Wire (2 mm diameter)

A motor was used to drive the belt and pulley mechanism. The speed of crosshead was controlled by syncing motor rotations with crosshead motion. For one rotation of motor, displacement of crosshead was measured; using time to make this displacement, speed of crosshead was calculated. For strain measurement, change in length was measured by counting how many rotations lead screw made for one rotation of motor and dividing this value by gauge length of specimen. The load applied was measured by using load cell. All these calculations were performed in MATLAB and stress–strain graph is plotted on GUI (Graphical user interface) using the data accusation by Arduino. Prototype with labeled parts are shown in Figure 2.8.

2.6 ENVIRONMENTAL IMPACT

The two categories in which UTM has environmental impact are as follows: design and manufacturing of the machine, and its operation. The waste material produced during manufacturing is swarf (metal chips) from machining of mechanical parts such as crosshead, base, etc. Swarf does not have much impact on the environment because it can be recycled by melting but there is some CO_2 emission associated with recycling. Special equipment is available for recycling which convert scrap to briquettes (metal cylinders). These mechanical parts are made of steel and carbon emission factor for recycling is 0.44 kg CO_2/kg steel [5]. So, recycling is producing less carbon dioxide which is a good factor for environment. These briquettes can be used again in manufacturing or it can be sold separately as a by-product.

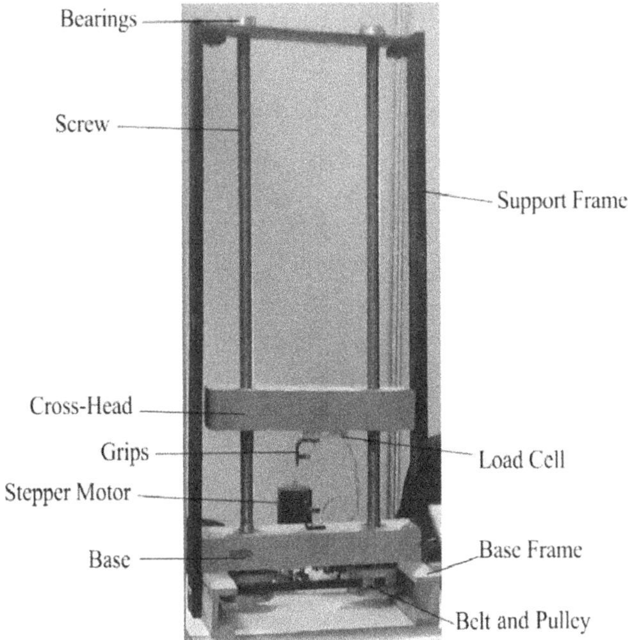

FIGURE 2.8 Design of prototype (labeled).

The other carbon emission is associated with design of machine. This includes emission related to material used in machine like steel, copper, plastics, aluminum, etc. The emissions in this respect are calculated with kg of CO_2 produced per kg of respective material values. CO_2 emissions are tabulated in Table 2.2.

Total carbon emissions in design of machine were calculated to be about 662 kg CO_2.

The environmental impact is due to operating equipment like motor, computers, and microcontrollers, etc., which include power consumption. This machine uses stepper motors which consume low power and provide better control. Conventional machines use servo motors with high torque capacity which consumes more power. The stepper motor used in this design was coupled with a gear box to increase torque capacity while keeping power consumption as low as possible. A quantitative comparison of power consumption between this design and an available model in market, i.e., UTEST's UTM-8050 [6], is shown in Table 2.3.

TABLE 2.2
CO_2 Emission Values

Material	Mass (kg)	CO_2 Emission (kg of CO_2)
Steel	418.6	650.923
Copper	0.2	0.542
Plastic	0.02	0.0372
Aluminum	1	11.46

TABLE 2.3
Comparison of Power Consumption of the Machines

Machine	Motor Type	Power Consumption (W)
Designed Machine	Stepper	50
Present Machine	Servo	300

Designed machine consumes 250 W less power than that of an available model, i.e., UTEST's UTM-8050. Percentage reduction in power consumption is given as follows:

$$\text{Percentage Power Reduction} = \frac{300 - 50}{300} \times 100 = 83.33\%$$

As carbon emissions are directly related to power production and consumption [5], the designed machine can reduce the carbon emissions due to power consumptions (during the operation) by the same percentage. Hence the devised machine design is also an environment-friendly design.

2.7 CONCLUSION

Functional reverse engineering approach was implemented in designing the machine according to required specifications and the resulting design improvements in UTM have been presented in this chapter. The machine was designed to have a better position control (given the use of stepper motor and gear box) and low CO_2 emissions during its operation reasoning to lower electrical power requirements of stepper motor. Mechanical design along with FEA showing robust design of critical parts is also presented. Furthermore, the electronic circuit has been detailed and validated by its successful implementation on the developed prototype. This machine demands low manufacturing and operational costs. This work provides a guideline on the use of functional reverse engineering approach in design of UTM. Further research can be carried out in designing GUI, improving electronics circuitry, and making the machine light weight.

ACKNOWLEDGMENTS

This work was supported by Directorate of Science and Technology (DOST), Government of Khyber Pakhtunkhwa (Pakistan), through a project titled: Promoting Enterprises of Reverse Engineering (Grant no. Dirtt/S&T/KP/PLC/ADP/2016–17).

REFERENCES

1. M. Fridman, "Quality Magazine," BNP Media, 1 April 2017. [Online]. Available: https://www.qualitymag.com/articles/93903-the-universality-of-a-universal-testing-machine. [Accessed 1 December 2018].
2. "The Universal Grip Company," UGC, 2019. [Online]. Available: https://www.universalgripco.com/utm-introduction. [Accessed 23 May 2019].

3. "Tinius Olsen," Tinius Olsen, [Online]. Available: https://www.tiniusolsen.com/our-company/history. [Accessed 23 May 2019].
4. "ASTM E8 / E8M-16a, Standard Test Methods for Tension Testing of Metallic Materials," ASTM International, West Conshohocken, 2016.
5. G. Hammond and C. Jones, *Inventory of Carbon & Energy (ICE)*, University of Bath, Bath, 2011.
6. "Utest Material Testing Equipment," Utest Material Testing Equipment, [Online]. Available: http://www.utest.com.tr/en/23649/Electromechanical-Universal-Test-Machine. [Accessed 24 May 2019].

3 Design, Modeling, Analysis, and Characterization of a Pin-on-Disk Tribometer with a Novel Spring Loading Mechanism

Ghulam Hussain, Shehroze Faisal, Iftikhar Ahmad, Wasim Ahmed Khan, M. Kamran Abbasi, K. Rehman, Sarmad Chohan, Ghulam Abbas, and Adeem Samad
GIK Institute of Engineering Sciences and Technology

CONTENTS

3.1 INTRODUCTION

The term Tribology is derived from the Greek word "Tribo" whose English translation is "to rub." Tribology is, therefore, the science of interactive surfaces in relative motion. The main areas of research and application are friction, wear, and lubrication. Friction can be productive for example in braking mechanisms, clutches, driving wheels, etc., and unproductive, such as inside internal combustion engines, gears, camshafts, etc. In the sliding and rolling surfaces of modern machinery, tribology is very essential to the design engineer as the knowledge of friction can greatly help in maximizing efficiency and economics [1].

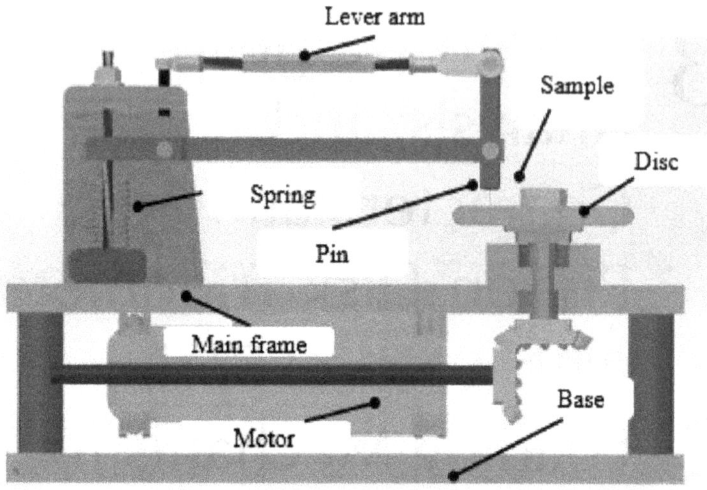

FIGURE 3.1 CAD model of the pin on disk tribometer.

The advances in the fields of materials and manufacturing technology have led to an influx of new materials that need to be tested with regard to their tribological behavior to determine their suitability to various operating conditions and environment [2]. A tribometer is a machine or a device used to perform these tests and simulations of wear, friction, and lubrication [3]. Often tribometers are extremely specific in their function and are fabricated by manufacturers who desire to test and analyze the long-term performance of their products [4]. Pin on disk tribometer is one such type of tribometer used to measure the friction coefficient. Although research and work are being done in the field of materials engineering in Pakistan, no appreciable work has been done so far in the field of tribology due to the lack of local manufacturers for the fabrication of devices like the tribometer. The ones manufactured internationally are available but too expensive for developing countries to afford. The demand for tribometer is considerable given the number of research facilities and institutes throughout the country which is entering the fields of materials engineering [5,6].

In this work, an economical yet high-precision Pin-On-Disc tribometer has been designed, developed, and successfully tested. The tribometer developed with a novel spring-loading mechanism can measure both static and dynamic coefficients of friction between any two surfaces. The CAD model of the developed Pin-On-Disc tribometer is presented in Figure 3.1.

3.2 DESIGN AND CONFIGURATION

The apparatus was originally designed using the ASTM standard for Pin on Disk testing (ASTM G99) as a guideline for functional requirements [7]. This specification is broad concerning the actual configuration of the apparatus, constraining only the system-level requirements (i.e., specific pin-to-disk interface and alignment requirements, relative motion, etc.). Major efforts were to focus on a cost-effective solution that would provide consistent and reliable results. As directly specified in ASTM G99 Section 3.1, the Pin-on-Disk wear test consists of two test specimens; a pin with a radioed or flat tip which is positioned perpendicular to the other specimen, usually a flat circular disk. The test machine causes either the disk specimen or the pin specimen to revolve around the disk center. In either case, the sliding path is a circle on the disk surface. The plane of the disk may be oriented either horizontally or vertically. The pin specimen is pressed against the disk at a specified load, often by means of an arm or lever. The load is then applied to the pin specimen (normal to the face of the disk) and maintained throughout the test [8].

The test apparatus designed for this project was configured with a load arm/disk spindle-type configuration. The test disk component was fastened directly to a cylindrical "spindle" using a shoulder screw to ensure accurate alignment and concentricity between the test disk and the spindle. The bearings provided the supporting reaction force against thrust loads applied normal to the disk test surface and also allowed rotation of the spindle/disk assembly about the axis of the spindle. Ball bearings were selected over roller thrust bearings for this application as they introduce less frictional drag force into the system. Future testing with this apparatus could potentially incorporate fluid contaminants so sealed bearing units were selected (also effective in preventing the ingress of pin wear debris) [9]. The section view of the developed tribometer is shown in Figure 3.2.

A 90° miter gear system was used to transmit the power directly from the gear motor to the spindle which provides the relative motion between the (rotating) test disk and the static test pin (loaded from above). The miter gear mounted on the output shaft of the gear motor was manufactured from case-hardened alloy steel while the mating gear (mounted to the lower shouldered end of the spindle) was manufactured from a brass alloy.

The Test Pin specimens were contained in an upper load arm assembly, consisting of a number of components which are illustrated in Figures 3.3 and 3.4. Initially, the test pins were installed in a "pin holder" component (blind hole, slip fit installation) and retained with a set screw mounted transversely. The pin holder is pinned at two locations in a four-bar-type linkage configuration which is used to ensure constant

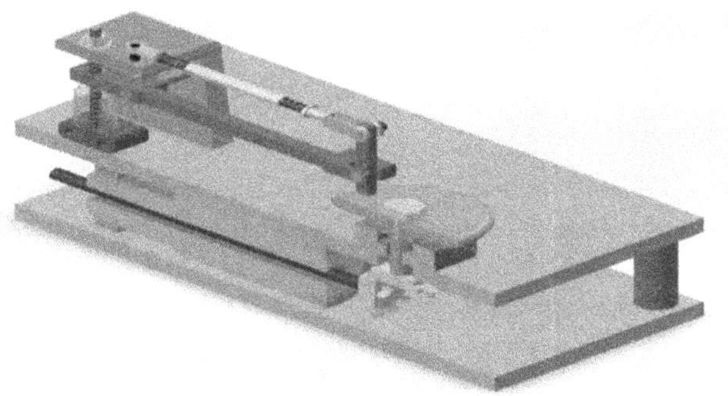

FIGURE 3.2 Section view of the pin-on-disc tribometer.

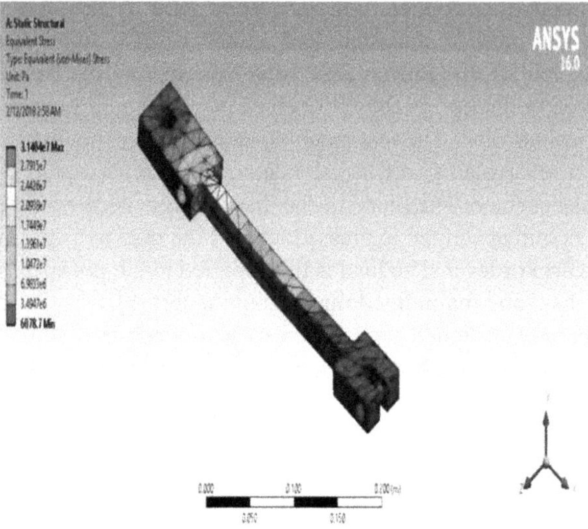

FIGURE 3.3 Stress distribution in load arm.

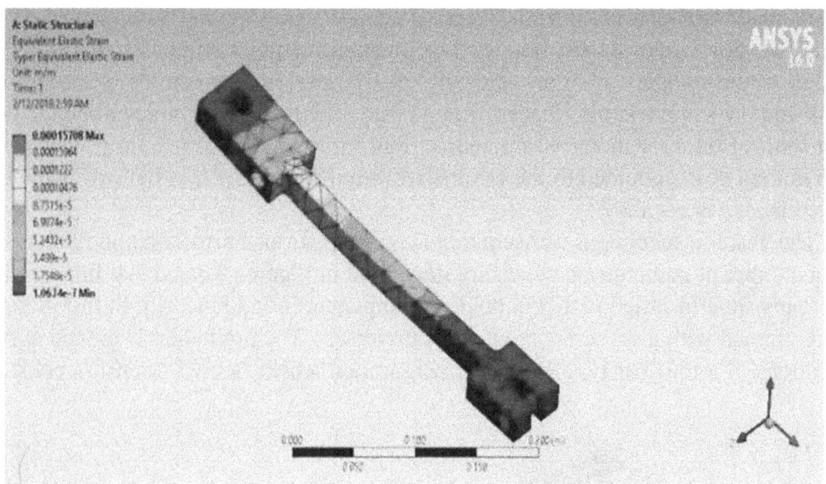

FIGURE 3.4 Strain distribution.

vertical alignment of the test pin. The pin holder was pinned at the uppermost location to a control arm assembly, consisting of five individual components. The control arm assembly utilizes an RH/LH threaded rod with a threaded clevis pin at one end (mates with the pin holder) and a female spherical bearing at the opposite end. By rotating the threaded rod, the operator changes the effective length of the control arm assembly, which in turn changes the angle of incidence between the pin holder and the mating disk surface. Once the operator has dialed in the correct length of the control arm and confirmed the absolute vertical alignment of the pin holder, the length is locked down using LH/RH jam nuts. This configuration guarantees that as the pin holder rotates about the pinned center of the load arm component (to which it is also attached), the test pin will always remain vertical and normal to the test surface [10]. The normal force of the pin on the disk F^N, the disk moment M^d, and the spring force F^s along with their direction is depicted in Figure 3.6.

The load arm component is central to the operation of the entire test assembly. The load arm is pinned at two locations: the outermost end is machined with a close-tolerance clevis/through-hole configuration to allow installation of the pin holder and retention with a spring pin. A larger vertical through-hole is located outboard of the pivot pin location; this hole allows sufficient clearance for a threaded lead screw which centers through a compression spring [11].

While evaluating all of the commercially available Pin-on-Disk Tribometer solutions on the market, it was noticed that in all cases, the normal force test load limit was relatively low (max load ranging 10–60N). It was this requirement that ultimately led to the selection of a compression spring-driven load design. Other load application methods that were considered during the design phase were a calibrated (slung) deadweight configuration and also the use of hydraulic/pneumatic actuators; however, the spring design offered the best combination of adaptability, control, and reliability at a significantly reduced cost. This test assembly uses an actuator that compresses the spring producing an upward force. The upward force on the load arm creates the FN test load on the test pin which is installed at the opposite end of the load arm. A close tolerance RC fit between the pivot pin and the load arm ensures that the test pin remains static relative to the rotating test disk. Dry film lubricants were used to mitigate wear between the pivot pin and the load arm.

3.3 ANALYSIS

The analysis was performed using ANSYS static structure to assess the stresses in the key component that is the load-bearing arm. After applying the appropriate boundary conditions to get accurate simulations to find the critical stresses and strains and hence the maximum force that can be applied within the safety limits. Analytical calculations were also carried out to compare with computational results. Aluminum 7075-T651 alloy was selected as the material for load-bearing arm due to its high tensile strength, lightweight, economic, and manufacturing ease. The yield strength for this alloy is 503MPa and has a modulus of elasticity of 71.7GPa. Using a safety factor of 2.5 the maximum operating stress is 202MPa. When a load of 540 Newton was applied a resultant maximum stress of 31.5MPa was computed in ANSYS as can be seen in Figures 3.3–3.5. This stress is well below the maximum operating stress.

Strain analysis was carried out to find the most suitable locations for the placement of strain gauges. In regions of no to little strain, there will not be enough deflection so strain gauges will need to be extremely sensitive which can lead to errors. Similarly, in regions of very high strain, there will be too much deflection so a little error will be amplified considerably. The strain gauges will therefore be placed in the region of moderate strain to get better and more accurate readings.

The most important aspect of the load-bearing element is the deformation and the detailed study which was necessary for the system to adhere to the set requirements.

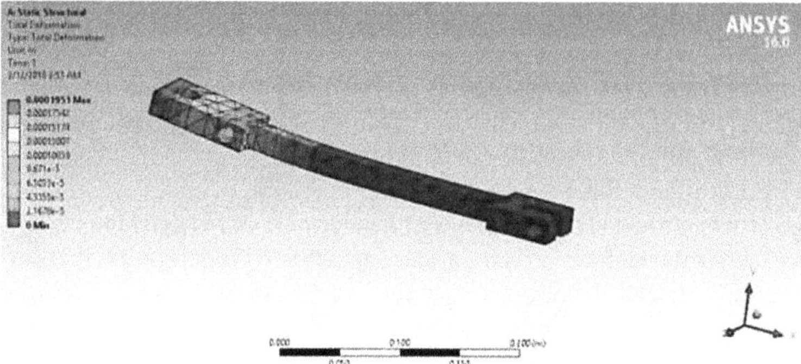

FIGURE 3.5 Deformation of the load arm.

The load-bearing arm is responsible for the application of the normal force on the pin since it is a pivoted assembly, and force is being applied on one end and the other end holds the pin which is constrained by the four-bar mechanism.

The maximum deformation is 0.19 mm for a load of 540 Newton. As can be seen in Figures 3.3–3.5, the maximum deformation occurs at the point of application of the force, i.e., the contact point of spring and load arm. It can also be seen that deformation occurs in the neck of the load arm and thus is responsible for the transfer of force to the pin. This force will be the normal force and used to find the coefficient of friction.

3.4 GOVERNING EQUATIONS AND MATHEMATICAL MODELING

The input force on the loading arm produces moment on the loading arm and it tries to bend the arm in a radius of curvature of the neutral axis [12].

$$\frac{1}{\rho} = \frac{M_{(x)}}{EI} \tag{3.1}$$

ρ = the radius of curvature of the neutral axis

$M_{(x)}$ = bending moment varies with respect to distance "x" from one end of the beam

E = Young's modulus

I = moment of inertia of the cross-section about the neutral axis

$$\frac{d^2y}{dx^2} = \frac{M_{(x)}}{EI} \tag{3.2}$$

Equation (3.2) is a second-order linear differential equation. It is the governing equation for the elastic curve. The product EI is known as the "flexural rigidity."

$$(EI)\frac{dy}{dx} = \int_0^x M_{(x)}\, dx + C_1 \tag{3.3}$$

If the flexural rigidity varies along the beam, as in the case of a beam of varying depth, it needs to be expressed as a function of "x" before proceeding to the integration of Equation (3.3). However, in the case of a prismatic beam, which is the case assumed during our analysis, the flexural rigidity is constant. The integration of Equation (3.2) results in Equation (3.3). Equation (3.3) is used to determine the slope of the tangent to the elastic curve.

$$\frac{dy}{dx} = \tan(\theta) = \sin(\theta) = \theta_{(x)} \tag{3.4}$$

The angle "$\theta_{(x)}$" is the angle at which the tangent to the elastic curve forms with the horizontal axis of the beam at any point on the elastic curve. It is measured in radians. Equation (3.4) is valid only for small values of angle θ.

$$(EI)y = \int_0^x dx \int_0^x M_{(x)}\, dx + C_1 x + C_2 \tag{3.5}$$

The further integration of Equation (3.3) results in Equation (3.5). Equation (3.5) is used to determine the deflection of the elastic curve at any point on the beam. The constants of integration "$c1$" and "$c2$" are determined from the boundary conditions. At the fixed supports, the deflection "y" is always zero. The maximum deflection "$y_{\max}$" in the beam occurs at the location where slope $\theta_{(x)} = \frac{dy}{dx} = 0$.

$$F = kx \tag{3.6}$$

The spring constant is a measure of how stiff the spring is. The larger the spring constant is; the more force is required to stretch or compress it. If the spring is forced to compress, it will resist that compression and try to restore itself to its equilibrium position. We can measure this by the simple equation known as Hooke's law. Equation (3.6) is the mathematical representation of Hook's law [13].

F = force applied to the spring in Newton

K = spring constant measured in Newton per meter

X = distance the spring is compressed from its equilibrium position in meters

$$f = \mu N \tag{3.7}$$

F = frictional force measured in Newton

N = normal force measured in Newton

M = coefficient of friction, a dimensionless number, determines the amount of friction. Its value depends upon the nature of the surface and type of the material.

3.5 PHYSICAL MODEL DEVELOPMENT AND TESTING

The tribometer was manufactured in-house in the faculty workshop. The detailed list of required materials along with the complete drawings was prepared with finalized dimensions. The structure of the main table was made from ASTM A36 Steel

FIGURE 3.6 Fabricated model.

because it was rigid and would not deform due to the weight of the components and spring force. The material was procured from the local market. The bearings were also purchased from the local factory inventory. The special grade aluminum that was used was procured from a local manufacturer. Similarly, the thrust ball bearing was also purchased from the same source.

For the joining purposes, nuts and bolts were used. For alignment of the supporting blocks, dowel pin was used in addition to nuts and bolts to ensure the alignment of the structure. After the assembly was complete the machine was checked for any defects or misalignment. When fully satisfied with the machine build, surface finishing was applied. For strain measurement, HBM 350 Ω strain gauges were used in the load arm. They were placed at almost the center of the load arm. These sensors provide a resistance output that is connected to a quarter bridge to convert resistance into a voltage which is then amplified using amplifiers. An Arduino microcontroller was used to record the data. The physical model of the machine along with the data acquisition system is shown in Figure 3.6.

3.6 RESULTS AND ANALYSIS

The basic purpose of the pin on disk tribometer is to determine the coefficient of friction of different materials. Several materials were tested using the tribometer and their coefficients of frictions were plotted against time. There was an option to use either the pin as a sample material or the disk. We used both the options while testing and results were similar.

First, a pin and a disk both made of mild steel were used during material testing using the tribometer. An input load was applied through the pin on disk. The disk was rotated at a particular rpm. The resulting frictional and normal forces generated on the pin due to contact with the rotating disk produced deflections on the loading arm which were measured using strain gauges through quarter strain bridge circuit and the output voltage was then amplified by the amplifier before feeding it into the Arduino. The graph of the coefficient of friction of mild steel vs mild steel plotted

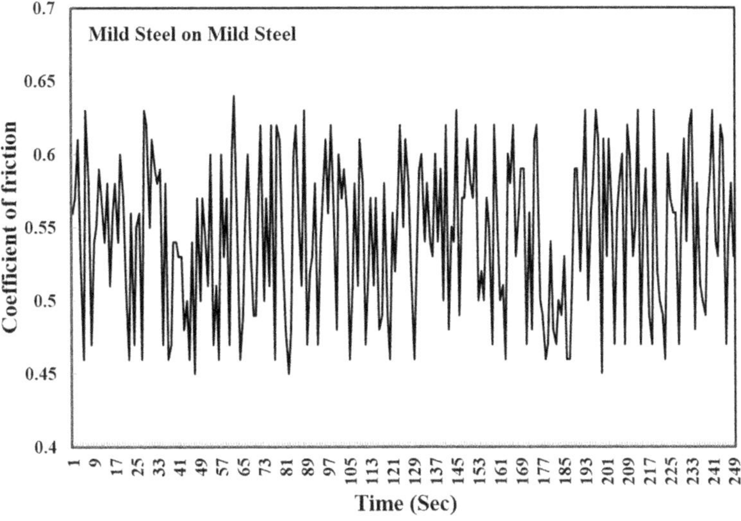

FIGURE 3.7 The coefficient of friction of Mild steel against Steel as a counter body.

is shown in Figure 3.7. The variations and fluctuations in the graph about the mean value are due to the vibrational effects of the pin. The mean value of the coefficient of friction for this set of materials is found to be 0.54 which is very close to the original value. The same experiment was performed under different loading conditions and the results of the coefficient of friction were similar for the same set of materials.

In the second case, the pin of mild steel and disk of aluminum were used during the testing. The graph of the coefficient of friction of mild steel vs aluminum plotted is shown in Figure 3.8. The value of the coefficient of friction fluctuates about its

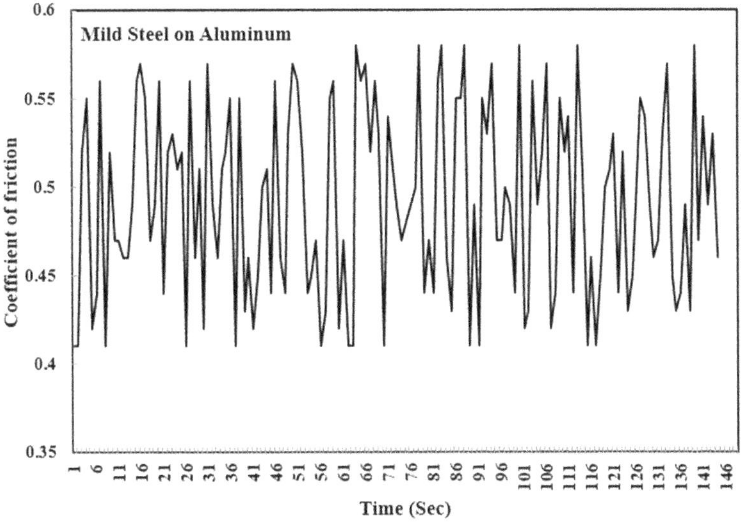

FIGURE 3.8 The coefficient of friction of aluminum against Steel as a counter body.

TABLE 3.1
Comparison of the Coefficient of Friction Values

Material	The Actual Value of the Coefficient of Friction	The Experimental Value of the Coefficient of Friction	% Age Difference
Mild steel vs. mild steel	0.57	0.54	5.3
Mild steel vs. aluminum	0.47	0.49	4.26

mean value. The mean value of the coefficient of friction for this pair of materials is found to be 0.49.

The actual and experimentally determined values of coefficient of friction of different materials along with the percent difference are mentioned in Table 3.1.

3.7 CONCLUSION

The basic purpose of the project was to introduce novelty in the design of the tribometer using a compressible spring mechanism. The tribometer was fabricated based on the proposed novel design to test the tribological properties of different materials. The coefficients of frictions of different sets of materials were found under testing using the tribometer. The sets of materials tested include mild steel vs mild steel and mild steel vs aluminum. The values of the coefficient of friction were experimentally found are quite close to the actual values. The experimental results differ slightly from the actual ones. These differences may occur due to the vibrations of the pin, external disturbances such as fluctuations in the applied voltage, pin not exactly normal to the surface of the disk, etc. The results obtained from the tribometer are satisfactory. The same tribometer can be used for other sets of materials as well and their coefficients of frictions are determined. If the material to be tested is costly or precious, there is an option to make the pin of that material and disk of other relatively cheap material. If this solution is not applicable when there is another option as well. A thin membrane or layer of the expensive material is coated over a cheap one to form a sample disk and thus the material properties of the expensive material can also be tested in a more economical way. The coefficient of friction is a material property, and it remains the same no matter the loading condition is as far as the pair of contacting materials is the same. However, the wear volume of the materials depends upon the loading conditions. The more the input force applied by the pin on disk or the more the rpm at which the disk is rotated; the more will be the volume loss due to wear. The more the contact area is the more will be the volume loss by wear.

REFERENCES

1. Marjanovic N, Tadic B, Ivkovic B, Mitrovic S. Design of modern concept tribometer with circular and reciprocating movement. *Tribology in Industry.* 2006;27(1&2):3–8.
2. Breki AD, Gvozdev AE, Kolmakov AG, Starikov NE, Provotorov DA, Sergeyev NN, Khonelidze DM. On friction of metallic materials with consideration for superplasticity phenomenon. *Inorganic Materials: Applied Research.* 2017 Jan 1;8(1):126–9.

3. Adamou AS, Denape J, Paris JY, Andrieu E. An environmental tribometer for the study of rubbing surface reactivity. *Wear.* 2006 Aug 30;261(3–4):311–7.
4. Hoić M, Hrgetić M, Deur J. Design of a pin-on-disc-type CNC tribometer including an automotive dry clutch application. *Mechatronics.* 2016 Dec 1;40:220–32.
5. Blau PJ. Elevated-temperature tribology of metallic materials. *Tribology International.* 2010 Jul 1;43(7):1203–8.
6. Li, D. *Static Coefficient of Friction Measurement using a Tribometer.* NANOVEA Application Notes, Irvine, CA, 2014.
7. Standard AS. G99. *Standard Test Method for Wear Testing with a Pin-on-Disk Apparatus.* ASTM International, West Conshohocken, PA, 2006.
8. Schmitz TL, Action JE, Ziegert JC, Sawyer WG. The difficulty of measuring low friction: Uncertainty analysis for friction coefficient measurements. *Journal of Tribology.* 2005 Jul 1;127(3):673–8.
9. Burris DL, Sawyer WG. Addressing practical challenges of low friction coefficient measurements. *Tribology Letters.* 2009 Jul 1;35(1):17–23.
10. Archard J. Contact and rubbing of flat surfaces. *Journal of Applied Physics.* 1953 Aug;24(8):981–8.
11. Stachowiak GW, Batchelor AW, Stolarski TA. *Engineering Tribology.* Elsevier, 1993, ISBN 0-444-89235-4, 960 pp, £ 156.00.
12. Ivanoff V. *Engineering Mechanics: An Introduction to Statics, Dynamics and Strength of Materials.* McGraw-Hill Higher Education, New York, 1995.
13. Khotimah SN, Viridi S. The dependence of the spring constant in the linear range on spring parameters. *Physics Education.* 2011 Sep;46(5):540.

4 Axes CNC Milling Machine

Sohaib Jabran, Shajee Ahmed, Sarmad Afzal, and Faizan Tariq
GIK Institute of Engineering Sciences and Technology

CONTENTS

4.1 INTRODUCTION

5-Axes machining constitutes three translational axes and two rotational axes. The manufacturing process of complex parts has been revolutionized by capabilities of multiaxes machining. Conventional machining processes are slow, inflexible, and limited in their functions and capabilities as compared to multiaxes machining such as 5-axes milling machine. Hence, by overcoming the constraints of traditional machining, it can provide the advantage of machining complex parts with increased accuracy and higher surface finish with reduced set-up time. Consequently, enhancing machine accuracy has been one of the main focuses of research on 5-axes machine tools in the past.

In design and manufacturing industry, NC (Numerical Control) concept refers to the automation of machine tools that are operated by abstractly programmed commands encoded on storage medium such as G-codes. The first NC machines were built in based on existing tools that were modified with motors that moved the controls to follow points fed into the system on paper tape. The early servomechanisms were rapidly improved with analog as well as digital computers, creating the modern computed numerically controlled (CNC) machine tools. A CNC machine includes several types: 2-axes CNC machine, 3-axes CNC machine, 4-axes CNC machine, 5-axes machine, etc. [1]. The number of the axis of a CNC machine implies capability of the controller of the machine to interpolate simultaneously. If the axis numbers are increased, the machining efficiency, effectiveness, and accuracy also increases; however, it requires more complex techniques in tool path generation and process of programming with advanced controllers such as raspberry pi. 5-axes milling CNC machine has been proven to be the most efficient tool for fabricating products of complex geometry which may include several free form surfaces. Integration of 5-axes CNC systems with CAD-CAM systems has revolutionized the industrial automation. The products are widely used in several high technology industries such as the aerospace industry, the automotive industry, the shipbuilding industry, etc.

4.2 LITERATURE REVIEW

Keywords—CNC (computer numerical controlled), GUI (graphical user interface), CAM (computer-aided manufacturing), NC (numerical controlled), MCU (machine control unit), read only memory (ROM), manual data input (MDI), random access memory (RAM), pulse width modulation (PWM).

The first NC milling machine was originated by John T. Parsons around 1940s–1950s. Parsons attached servomotors to the X- and Y-axis of a manually operated machine tool. Servo motors were connected to each axis to automate motion. These servo motors were programmed corresponding to the required shape. The reason for inventing such machine was to make complex shapes like arcs such as airfoils for airplanes. Traditionally, NC systems have been composed of the following components [2]:

Tape punch: converts written instructions into corresponding hole pattern. The pattern is punched into a tape.

Tape reader: it converts the hole pattern on a tape into electrical signals.

Controller: receives the electrical signal from tape reader and gives subsequent instructions to NC machine.

NC machine: it responds to the controller.

NC machines have an advantage over manual production systems:

1. Better control of tool motions under critical cutting conditions.
2. Better part quality and repeatability.

3. Reduces tool cost, tool wear, and set-up time.
4. Better production planning.
5. Reduces waste.

Today's modern technology is based on CNC milling machines and lathes. A CNC machine is an NC machine with the added feature of an on-board computer. This on-board computer is referred as MCU. In NC, MCU is "hard wired" that is all machine functions are controlled by physical electronic elements embedded in a controller whereas CNC is "soft wired." Thus, the machine functions are encoded into the computer at the time of manufacture. They will not be erased when CNC machine is powered off.

A microprocessor in each CNC machine reads the part program that the user creates in GUI and performs the programmed operations. Personal computers are used to design the parts and are also used to write programs by either manual typing of part program or using CAM software that outputs part program such as G-codes from the user inputs of cutters and tool path.

4.2.1 CNC Concepts

An important advancement in the philosophy of NC machine tools was the shift toward the use of computers instead of proprietary controller units in the NC system of the early 1970s. This gave rise to the CNC. CNC is a self-contained NC system for a single machine tool including a dedicated minicomputer controlled by stored instructions to perform some or all basic NC functions. These machine instructions or machine functions are all soft wired in CNC. Computer memory that holds such information is called ROM. The MCU has an alphanumeric keyboard for MDI of part programs. Such programs are saved in RAM. The data stored in ROM cannot be edited or deleted whereas in RAM the data can be changed or played back. When CNC machine is turned off, the data stored in RAM is lost. New CNC machines have advanced MCU units that have graphic screens to display CNC programs, cutter paths, and errors in the program. The combination of ROM, RAM, and computer instructions forms a microprocessor or single processing unit that distinguishes CNC from NC. Hence, it has become widely used for manufacturing systems mainly because of its flexibility and less time requirement.

4.2.2 Design Consideration of CNC Machine Tools

CNC machine tools must be better designed, constructed, and must be more accurate than conventional machine tools. It is necessary to minimize all noncutting machine time and minimize idle motions by increasing the rapid traverse velocities to make the use of the machine tool more efficient.

Digital control techniques and computers have undoubtedly contributed to better accuracy and higher productivity. However, it should be noted that it is the combined characteristics of the electronic design as well as the mechanical design of the machine tool itself that determine the final accuracy and productivity of the CNC machine tool system.

FIGURE 4.1 Ball screw.

High productivity and accuracy might be contradictory, and one needs to find a tradeoff between the two. Because high productivity requires higher feed, speed, and depth of cut, which increases the heat and cutting forces in the system. This will lead to higher deflections, thermal deformations and vibration of the machine, which results in accuracy deterioration. Therefore, to achieve high operating bandwidth while maintaining relatively high accuracy, the structure of CNC machine tool must be more rigid and stiff than its conventional counterpart [3].

To achieve better stiffness and rigidity of structure, several factors should be considered in the design. The first concern is the material. Conventional machine tools are made of cast iron. However, the structures of CNC machines are usually all-steel-welded, constructed to achieve greater strength and rigidity for given weight. In addition, better accuracy is obtained in CNC machines by using low friction moving parts, avoiding lost motions and isolating thermal sources. Regular sliding guides have higher static friction than the sliding friction. The force used to overcome the static friction grows too large when the guide starts to move. Due to inertia of the slide the position goes beyond the controlled position, adding overshoot and phase lag to the system response, and affects the accuracy and surface finish of the part. This can be avoided by using slides and lead screws in which the static friction is lower than the sliding friction. In this machine, a ball-bearing lead screw is being used as shown in Figure 4.1.

4.3 DESIGN AND ANALYSIS

4.3.1 Design Methodology

4.3.1.1 Mechanical

A vast research was carried out before the selection of mechanical design of 5-axes CNC milling machine. From some research and functional reverse engineering of

3-axes precision router present at the disposal, a deduction was made that all the linear axes will contain following components.

- Guide rails
- Ball screw
- Screw block
- End blocks

From the study of *precision router*, it was also concluded that the final design will either be gantry type or knee and column type; depending upon the operational considerations of the machine.

After finalizing the design and fixture type of the three axes, next step was to design two rotating axes. From very basic torque calculations, it was concluded that these two axes will experience extensive torques even for cutting aluminum. If the torques were to be held directly from the motors, a large motor would have been required.

To counter this problem, two separate gear boxes, according to the torque requirements were designed using standard module gears. But later, geared stepper motors were selected which gave the required gear ratios hence the required torque at the output.

4.3.1.2 Electronics and Software

In computer science field, Interpreter Design is the conversion of source code or high level English like language into Binary or machine language. In CNC technology, Interpreter is a program or software that converts high-level language that is mnemonics or G&M codes to machine language and executes the source code. There are three steps in writing an interpreter for CNC (Figure 4.2).

4.3.1.2.1 Command Identification

In this step interpreter reads all the characters such as G, M, X, Y, Z, I, J, K, etc. It saves the block such as N35 G70 in controller memory in ASCII format according to the format classification. After this step, each character or alphabet is read from the word which resides at a memory location or by pointing to that memory location. Each letter in the word is stored either as its binary equivalent or as its UNICODE equivalent. The interpreter compares the content or character one by one with allowable characters by pointing to every memory location. This is achieved by incrementing or decrementing counter register or memory locator. The loop continues until all the characters in the word are compared.

FIGURE 4.2 Sequence of command.

4.3.1.2.2 Command Formation

After identifying all the characters in a block, interpreter reads all the numerals. In word N35 G70, the numerals are 3, 5, 7, and 0. Each numeral consists of Binary or any UNICODE equivalent. Same steps are repeated as in command identification. Interpreter reads and stores all the numerals one by one. The last step is to concatenate the word that combines alphabets and numerals.

4.3.1.2.3 Command Execution

This last step is performed by an interpreter. In this stage, interpreter calculates the numbers of pulses to be sent to the controller in order to generate the required motion from motors. After calculating the pitch of lead or ball screw, the number of turns per revolution is determined. The number of pulses determines the execution time. The speed provided by the user determines the delay or time required to energize one coil (Figure 4.3).

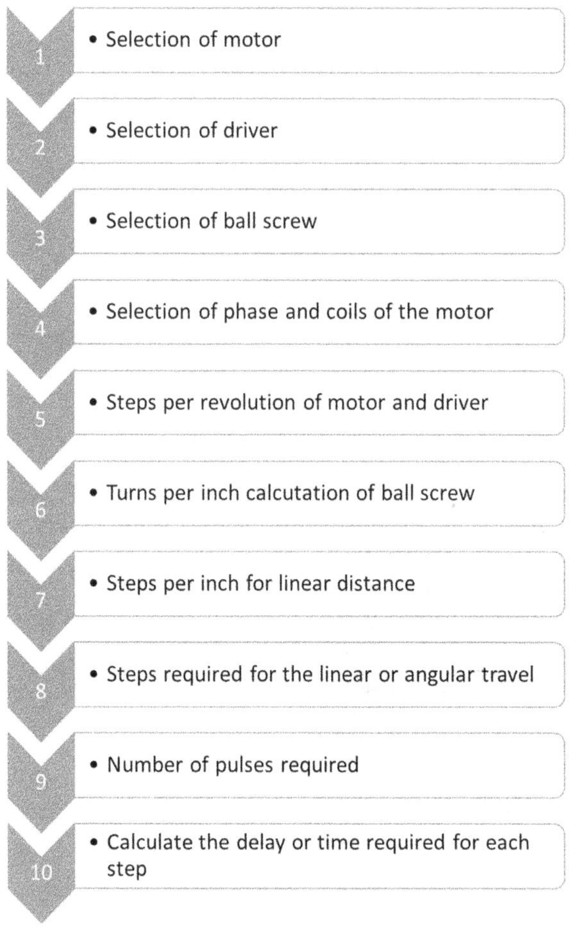

FIGURE 4.3 Command execution flow chart.

4.3.2 Governing Equations and Mathematical Modelling

4.3.2.1 Load Calculations

For X-Axis

Empirical equations for calculation of the cutting force [4]

$$F = Ct^x w \tag{4.1}$$

where

C: material constant at a given cutting speed and rake angle
t: uncut chip thickness in mm
w: width of the cut in mm or length of cutting edge engaged
x: constant for material being machined; it varies with rake angle

For this machine, the aluminum work piece was selected so the values of constant were following:

$C = 500$
$t = 0.15$
$x = 0.70$
$w = 20$

$$F_{cx} = Ct^x w$$

$F_{cx} = 500 \times 0.15^{-0.70} \times 20$
$F_{cx} = 2650\,\text{N}$
$F_{cx} = F_{cy}$; according to symmetry

$$F_{cz} = F_{cx} \tan(\beta)$$

$$F_{cz} = 2650 \times \tan(41.34) = 2330\,\text{N}$$

Hence the total force opposing the motion in X motor is given by

$$\text{Total load} = L = F_{cx} + \left(F_{cz} + F_{cy} + W_t\right) \times \mu_s \tag{4.2}$$

where $W_t = 200\,\text{N}$ (20 kg) is the total weight acting on the X-axis of the machine as shown in Figure 4.4.

The coefficient of friction is assumed to be 0.3
$L = 2650 + (2330 + 2650 + 200) \times 0.3$
$L = 4204\,\text{N}$

For Y-axis

With $W_t = 250\,\text{N}$ (25 kg) is the total weight acting on the Y-axis of the machine as shown in Figure 4.5, the coefficient of friction is assumed to be 0.3:
L = 2650 + (2330 + 2650 + 250) * 0.3
L = 4219 N

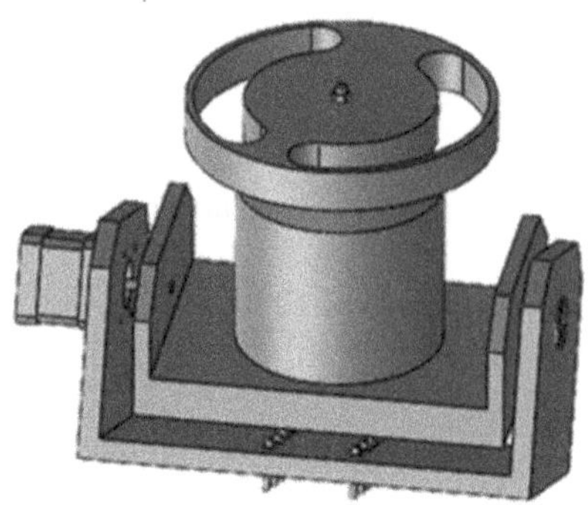

FIGURE 4.4 Load on X-axis.

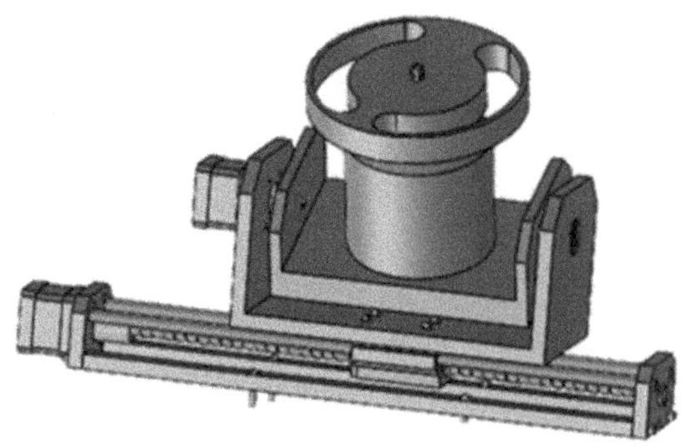

FIGURE 4.5 Load on Y-axis.

For Z-axis

$W_t = 50\,N$ (5 kg) is the total weight acting on the Z-axis of the machine as shown in Figure 4.6:

$$L = F_{cz} - W_t$$

$$L = 2330 - 50$$

$$\mathbf{L = 2280\ N}$$

4.3.2.2 Torque Calculation

For the calculation of the torque, following formula was used [5]:

$$T_r = \frac{d_m}{2} \times \frac{L(\mu \pi d_m + l)}{(\pi d_m - \mu l)} + L\mu_c \frac{d_c}{2} \tag{4.3}$$

where

$\mu = 0.07$

$d_m = 8\,\text{mm}$

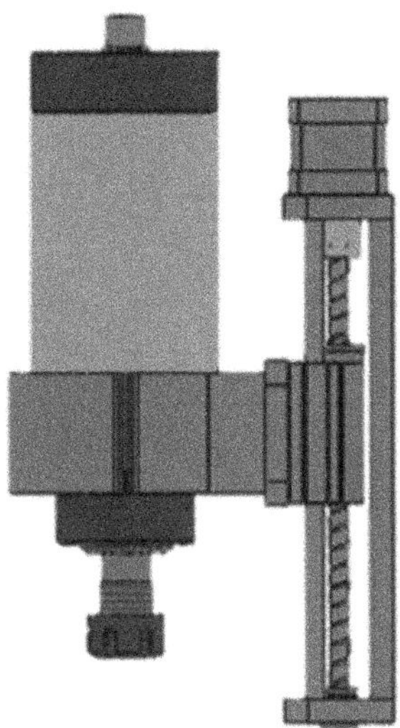

FIGURE 4.6 Load on Z-axis.

Pitch, l = 2.1 mm

Hence,

For X-Axis

$\mathbf{T_{r\ for\ x\text{-}axis} = 3.774\ Nm}$

For Y-Axis

$\mathbf{T_{r\ for\ y\text{-}axis} = 3.788\ Nm}$

For Z-Axis

$\mathbf{T_{r\ for\ z\text{-}axis} = 2.047\ Nm}$

For A-Axis

This torque calculated for the motor which rotates the work part and holds the assembly about X-axis is called A-axis. For this rotation of the work part the equation used is

$$T_A = r \times F \tag{4.4}$$

Forces that contribute to the moment to be held by motor are as follows:

1. Weight (100 N) of the assembly shown in Figure 4.7
2. The cutting in Z-direction F_z (2330 N)

$$T_A = 100 \times (0.075)/\left(2 \times 2^{1/2}\right) + 2330 \times \left(0.075/2^{1/2}\right)$$

$$\mathbf{T_A = 126.218\ Nm}$$

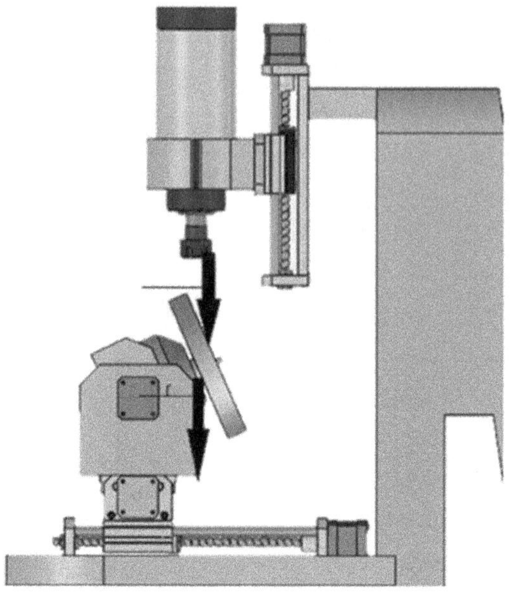

FIGURE 4.7 Load on A-axis.

For C-Axis

This torque calculated for the motor which will rotate the work part and hold the assembly about Z-axis is called C-axis (Figure 4.8). For this rotation of the work part the equation used is:

$$T_C = r \times F$$

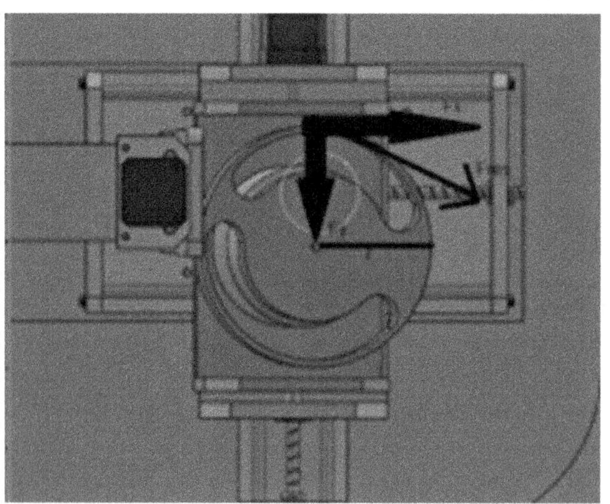

FIGURE 4.8 Load on C-axis.

Force that contributes to the moment to be held by motor is the cutting force resultant of F_x and F_y (3747.66 N).

$$T_C = 3747.55 * 0.05$$

$$\mathbf{T_C = 187.38\ Nm}$$

4.3.3 Geometric Modeling and Design

In the computer aided design process, several concept designs were made and after careful considerations, discarded. The concept designs included a planar body design in which *X*, *Y*, and *Z* axes are mounted on a gantry structure. Analysis showed there to be excessive vibration when cutting load was applied. Other models included column structure and knee and base structure, which failed the stability test.

The detail of rejected concept designs is excluded here to keep the paper concise and to the point.

4.3.3.1 Final Design

With the issues of bending in linear rails during the analysis part, the design is changed to linear rails and guides which are free from bending and apart from this, the base and back have been divided into two parts which will be casted separately.

Removal of support columns further reduced vibrations and increased stability of the final design (Figure 4.9).

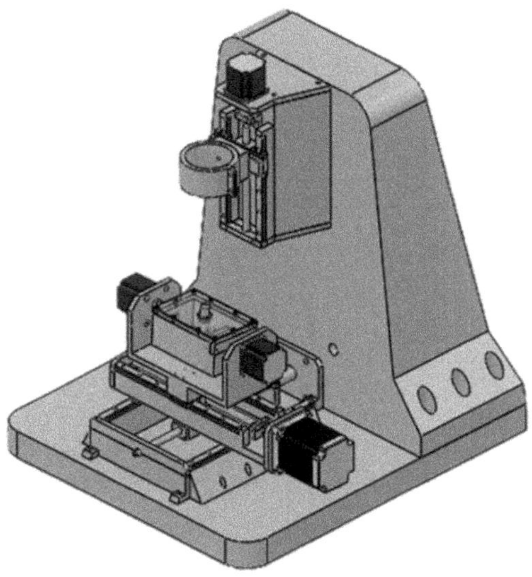

FIGURE 4.9 Finalized design.

4.3.4 Analysis

Most critical parts in the machine which were most susceptible to failure and could cause whole machine to fail were analyzed using ANSYS (Figure 4.10–4.12).

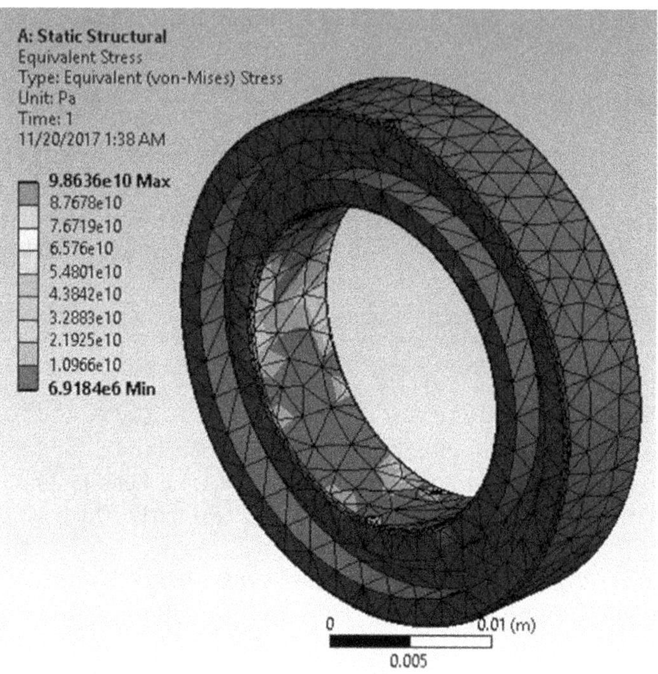

FIGURE 4.10 Equivalent Von Mises stresses on bearing.

FIGURE 4.11 Total deformation on bearing.

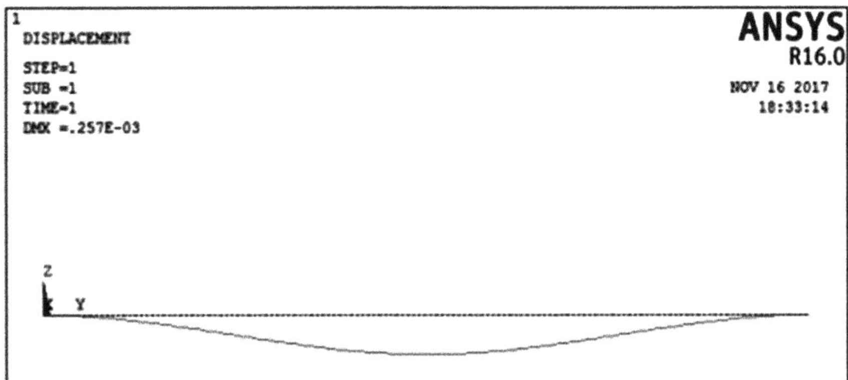

FIGURE 4.12 Guides-bending deformation.

4.3.4.1 Bearings

4.3.4.2 Linear Guides

Due to the shown deformation of 0.257 mm in the linear rails, the structure of the support may bend which will result in out-of-dimension part machined. Considering such consequences, the support assembly is changed from linear circular simply supported guides to linear rails which can easily support a 600 N load.

4.4 PHYSICAL MODEL DEVELOPMENT AND TESTING

4.4.1 Development Process

The development process of 5-Axes CNC milling machine involves development of software to run G-codes; the electronic circuitry to control the motors such that axes will move as given by user and interpreted in pulses of voltage; lastly, the mechanical part which includes first design, fabrication, and then assembly.

For the mechanical parts, the base and the column head will be casted of cast iron and accessories will be machined using material ASTM 36 structural steel. Furthermore, most of the parts have been kept standard to promote standardization and ease in maintenance of the machinery.

All the standard parts and machined parts are then assembled in the following manner:

1. Place base and connect the head using the nuts and bolts
2. Fasten the Y-axis rails and then assemble the rails over it for both sides
3. Join the end mounts and NEMA mounts of Y-axis and place the ball screw in middle using bearing on one end and connection coupling on the other
4. Connect the Y-axis connection plate using screws
5. Place the linear guides of X-axis over the Y-axis connection plate and connect then securely using bolts
6. Fasten the X-axis rails and then assemble the rails over it for both sides

7. Join the end mounts and NEMA mounts of X-axis and place the ball screw in middle using bearing on one end and connection coupling on the other
8. Fasten the rotation table, that is, X-axis connection plate.
9. Place the gears in rotation box and hang it from both ends using key and shafts
10. Mesh the bevel gears and ensure smooth motion
11. Seal the rotation box with cover on top and screw it
12. Fasten the Z-axis rails carts and then assemble the rails over it for both sides
13. Join the end mounts and NEMA mounts of Z-axis and place the ball screw in middle using bearing on one end and connection coupling on the other
14. Connect the Z-axis connection plate using screws
15. Place the motor holder on Z-axis connection plate
16. Place the spindle motor
17. Connect all the motors
18. Fasten shafts using couplings

4.4.2 Integration and Instrumentation

The components that are used in interfacing and integration are as follows:

1. Raspberry Pi
2. L298 H-Bridge Motor Driver
3. NEMA Stepper Motor
4. AC to DC Power supply

Raspberry Pi Model 3B is used as the microcontroller to control the motors for each axis. The input is given to Raspberry Pi in Desktop mode. The variables that are given as inputs are distance, speed, direction, and mode of interpolation that is circular or linear.

The motor selected for each axis depends on the cutting torque calculated above. The speed of motor is limited to the voltage supplied and also the torque. For X-, A-, C-, and Y-axes NEMA 34 is used and for Z-axis NEMA 23 is used. The power supply used for all motors is 12 V AC to DC but with different currents. For NEMA 34, 6 A current is used and for NEMA 23, 3 A current is provided. The mode is bipolar which provides more torque with minimum current and voltage (Figure 4.13).

4.5 CONCLUSION

4.5.1 Summary

This project was a prototype of the 5-axes CNC milling machine and it can be scaled up to the industrial scale which is the long-term goal of this project. The milling machine has uncountable applications in the industry. This will give a huge head start for the future projects in this regard.

The literature review was done right at the beginning of the project to get an understanding of what is happening in this regard in different parts of the world.

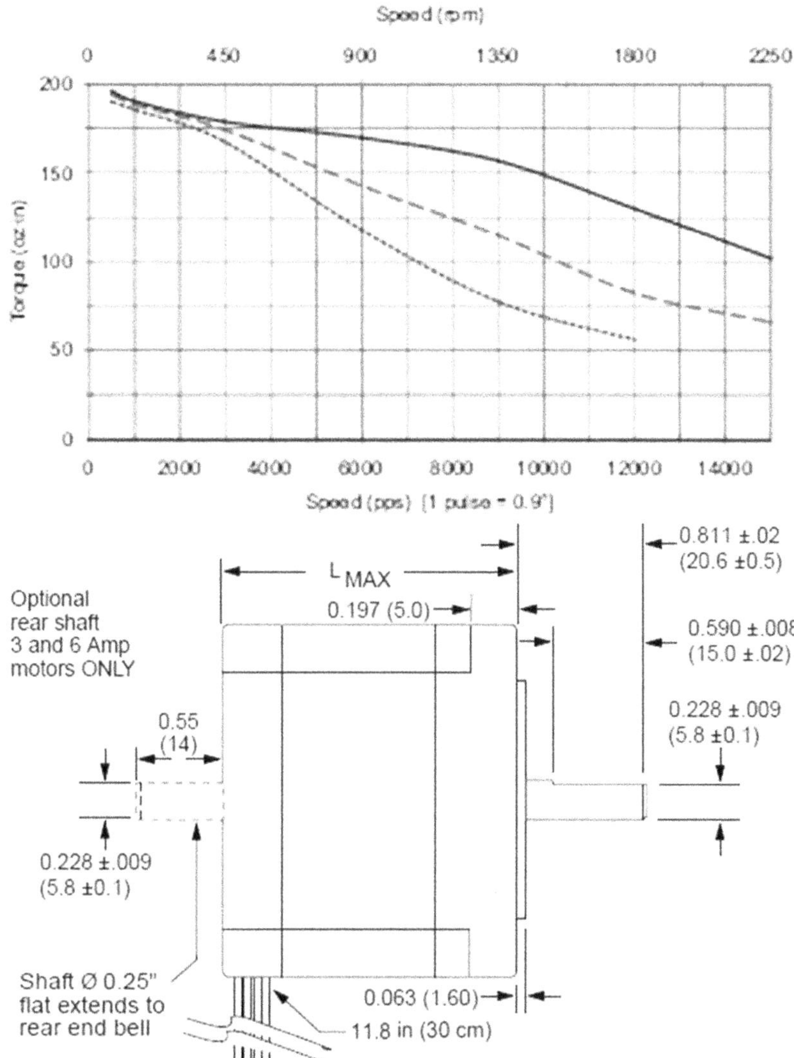

FIGURE 4.13 Motor specifications.

After that, the design methodology was adopted using the findings of literature review. Almost all the parts that were used in the machine were standard parts and available in the market; this excludes the prolonged procedure of manufacturing individual parts.

Simultaneously, the software part was also being coded. The language used to code the microcontroller Raspberry Pi was Python. The software was made by applying the systematic steps of format classification, command identification, command formation, and command execution.

4.5.2 Future Recommendations

The 5-axes CNC milling machine currently developed is a prototype machine with a work part box volume of $5 \times 5 \times 5\,cm^3$. This machine can be scaled up to design an industrial scale 5-Axes CNC milling machine which could perform heavy duty metal cutting of parts with intricacy and precision.

Furthermore, a machine always has room for improvements which in this case is the accuracy of the machine for precise cutting of complex shapes. Using better microcontroller stepping of the machine could be further divided to achieve demanded accuracy.

A business development plan can be developed to sell this machine is the market, since 5-Axes CNC milling machines are not easily available in the market despite its ability to machine very complex 3D shapes.

ACKNOWLEDGMENT

We wish to express our greatest of gratitude to Prof. Dr. Wasim Ahmed Khan for providing us with this wonderful opportunity. We went out of our ways to make sure that this project gets to the point where it deserves to be. Members whole heartedly dedicated themselves toward the cause and all these have only happened due to the sincere efforts of Dr. Wasim.

Moreover, we would like to thank Dr. Khalid Rehman for playing a vital role in the timely completion of the report. Lastly, we would also like to thank all the references to grant us the permission to quote their brilliant pieces of work. We are thankful to all those who are in any way responsible for creation and compilation of this paper.

REFERENCES

1. The What, Why and How of 5-Axes CNC Machining (ND). From https://www.engineering.com/BIM/ArticleID/11930/The-What-Why-and-How-of-5-Axis-CNC-Machining.aspx
2. James V. Valentino and Joseph Goldenberg (2003), *Introduction to Computer Numerical Control (CNC)*, 3rd edition, London, UK: Pearson Education.
3. Peter Smid (2003), *CNC Programming Handbook*, 2nd edition, New York: Industrial Press, Inc.
4. Mikell P. Groover (2010), *Fundamentals of Modern Manufacturing*, 4th edition, New Jersey: Wiley. "Chapter No. 21: Material Removal Processes".
5. Richard G. Budynas and J. Kieth Nisbett (2006), *Shigley's Mechanical Engineering Design*, 8th edition, New York: McGraw Hill. "Chapter No. 8: Screw, Fasters and Design of Non-Permanent Joints" - Eq. # 8–1.

Section 2

Sensors, Transducers, Printed Circuit Boards, and Control

Currently the experimentation with machine tools development is based on selection of microcontrollers, drivers, motors, and electronics components. The machine tool developers, all over the world, rely on same strategy with an easy access to manufacturers of these items allowing them to have a component made on their machine tool/machine tool controllers specification.

As Pakistan is currently striving for manufacturing independence in many of these areas, an overdesign in many cases is accepted as the only option available currently. With due course of time, these items will also be produced with maximum possible deletion strategy if the economy of scale permits it.

This section deals with the selection of sensors, transducers, and actuators. The development of piezoelectric sensors is addressed. It also addresses the printed circuit boards development as well as the control engineering.

5 Sensors and Actuators
Selection and Interfacing

Memoon Sajid and Mazhar Javed
GIK Institute of Engineering Sciences and Technology

CONTENTS

5.1 INTRODUCTION

A modern mechatronic system is a set of mechanical, electrical, and computer modules connected together as presented in Figure 5.1. The selection of suitable sensors and actuators for a given system design is a crucial task to make the system operate in a desired manner. Many different parameters have to be considered for the selection of these devices.

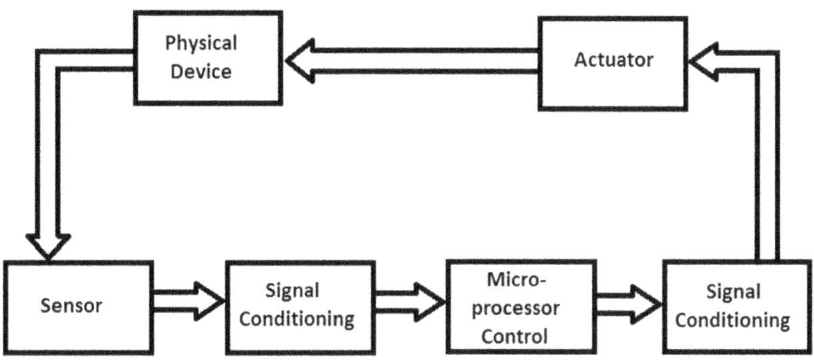

FIGURE 5.1 A simple microprocessor controlled mechatronic system.

There is a huge variety of sensors and actuators to choose from. They are divided into different categories based on structure,[1–3] size,[4] material,[5–7] principle of operation[8,9] type of signal involved,[10,11] and application.[12] Sensors we come across in our daily lives include temperature sensors,[13–15] pressure sensors, flex sensors,[16,17] light sensors,[10] humidity sensors,[18–20] touch sensors, current sensors, etc. There are different types of sensors including capacitive, resistive,[15] voltaic, amperometric, chemical, electrochemical,[11] colorimetric,[21] physical or mechanical,[16] and acoustic.[8] Different types of materials are used to fabricate these sensors including organic materials,[12] polymers,[22] in-organic materials, oxides, nanomaterials,[7,11] composites,[19,23] biomaterials,[9] etc. This huge variety of sensors is there to address different requirements of different systems and environments. For instance, humidity sensors made up of metal oxides are suitable for high-temperature environments, humidity sensors made from biocompatible materials are suitable for healthcare systems, polymer-based sensors are suitable for general environmental monitoring, and so on. Similarly, transducers are fundamental in converting a signal in one domain to another. A colorimetric transducer will convert an electrochemical reaction's products into a visible change in color that can be directly detected by the human but is difficult to be interpreted by a machine. A transducer targeted for machine-readable signal conversion will convert the same product of an electrochemical reaction into change in current, voltage, resistance, or any other electrical, mechanical, or optical property that can be read by the desired system. Transducers are usually there as an integral part of sensing devices. Sensors detect a certain analyte and the resulting change in the properties of sensors is actually converted to a human or machine-readable signal by the transducer. For example, polyvinyl alcohol is a polymeric material that is sensitive to the change in humidity. The dielectric constant that is a physical property of the material, changes with changing humidity. That change in dielectric constant cannot be measured/read/comprehended/detected without converting it to a signal that can be read or interpreted by any human or machine. An interdigitated transducer electrode is used to convert that change in dielectric constant into the change in capacitance. Similarly, in case of PEDOT: PSS based humidity sensor, the change in resistance of the material is converted into an electrical signal using the transducer electrodes. This means that the transducers usually do not need to be selected separately but they are already parts of sensors and the selection of sensors and designing of signal conditioning circuits for the sensors is based on the type of signal the transducer is providing as the output. Many types of actuators are also common including hydraulic actuators,[24] pneumatic actuators,[25] electrical actuators,[26] mechanical actuators,[27] thermal actuators, electrochemical actuators, shape memory alloys and polymers,[28] piezo-electric actuators, magnetic actuators, and smart actuators.[29] Different types have different properties and performance parameters and require different types of input signals to operate. A single system is usually designed to work in a known environment and in a defined range with desired accuracy. Most sensors, transducers, and actuators used in a single system usually belong to similar categories and involve similar signals at their inputs and outputs. This chapter will discuss in detail how to select these devices for a given system and how to interface them.

5.2 SHORTLISTING OF DEVICES

Sensors and actuators of hundreds of types and working principles are at our disposal these days as discussed in Chapter 2 of Ref. 30. It may become a very confusing and complicated process to select the right device for the targeted application and system. The selection of these devices is usually done in two major steps. First one is to shortlist a bunch of devices based on the required parameters and then finally select one device that is most suitable for the given problem at hand. Let us first discuss the process of shortlisting of these devices to a manageable number.

5.2.1 Sensors

A sensor is a device that when exposed to an analyte like temperature, pressure, water, bioreagents, etc., produces an output signal of its own through the integrated transducer in terms of change in resistance, capacitance, voltage, displacement, color, etc. The first step in selection of sensing devices is to determine the desired performance parameters and ranges of operation. The general performance parameters associated with a sensing device include selectivity, sensitivity, resolution (accuracy), detection range, stability, signal-to-noise ratio, operating temperature, hysteresis, transient response time, and lifetime. These general parameters are enough to initially select a bunch of sensors from hundreds and thousands of available options.

Example 5.2.1

A temperature sensor is required to be installed inside the boiler of a steam turbine. The temperature range that the sensor needs to detect is between 70°C and 400°C. The temperature needs to be detected with an accuracy of ±2°C while the transient response time should be less than 5 seconds. The lifespan of the sensor should be no less than 2 years with a maximum of 2% degradation in response during this period. What type of sensors can be used for this task?

Solution

The given information is enough for a system designer to at least select around 10–12 possible candidate devices that can get the job done. First of all, the most crucial parameter is considered that is in the range of detection (70°C–400°C in this case). With reference to Table 5.1, the types of devices best suited for the job are thermocouples (–200°C to 2500°C) and resistive temperature detectors (RTDs) (–200°C to 800°C). Sensors in both these categories have a lifespan of several years without degradation, and with suitable signal conditioning circuitry, they can give accuracy within ±2°C. The only parameter left is the transient response time that can be now used to simply reduce the list further. Once a list of devices that can be used has been compiled, the next consideration will be the other factors that will be used to finalize a single device in coming sections.

One important point to note here is not to go for an overkill of a device for a simple task. That can be explained through another example with much less information on the exact values of required parameters and more reasoning and critical thinking involved as engineers and system designers.

TABLE 5.1
Comparison of Different Categories of Temperature Sensors

Sensor Type	Thermistor	RTD	Thermocouple
Temperature range (typical)	−100°C to 300°C	−200°C to 800°C	−200°C to 2500°C
Accuracy (typical)	0.05°C–1.5°C	0.1°C–1°C	0.5°C–5°C
Long-term stability @ 100°C	0.2°C/year	0.05°C/year	Variable
Linearity	Exponential	Fairly linear	Non-linear
Power required	Constant voltage or current	Constant voltage or current	Self-powered
Response time	Fast 0.12–10 s	Generally slow 1–50 s	Fast 0.10–10 s
Susceptibility to electrical noise	Rarely susceptible High resistance only	Rarely susceptible	Susceptible/cold junction compensation
Cost	Low to moderate	High	Low

Example 5.2.2

A humidity sensor is to be selected for a commercial weather monitoring system where the hundreds of sensor nodes are to be spread across the city to provide information in real time. Explain the process of shortlisting the sensors for this application.

Solution

The given data in this problem do not have any values but rather provides a scenario-based information on the topic. The design engineer has to do some research and define the requirements first and then start the shortlisting process. For weather monitoring, the relative humidity ranges from 0% RH to 90% RH so we have the desired range of operation. The operating temperature of the devices will be the surrounding temperature of open air that usually varies between −20°C and 60°C. The desired accuracy and resolution in this case are normally ±2% RH and 1% RH, respectively, while the response time may be a couple of minutes. The expected lifetime of these devices is usually 3–5 years depending on the first-time cost and the running cost considerations while the degradation must be less than 2%–3% over this period. One can easily find hundreds of devices fulfilling these requirements in the market so that does not make the task any easier. But it is known that the sensor nodes are mass deployed and for a commercial setup, the cost must be low. What a system designer can do here is, to first filter the devices with their best parameters near to the requirements and omit those devices from the list that far exceed the requirements at hand as they would be an overkill for this project. Second, the design engineer can sort the devices by their price and choose the cheapest one that just serves the purpose. One thing to keep in mind is that the second method may not be the ideal as the engineer may end with a device with perfect performance characteristics but not compatible with the rest of the system (signal conditioning and device output). More details on this issue will be discussed in later sections of this chapter.

5.2.2 Actuators

An actuator is a device that when given an input signal of a certain form, produces a physical mechanical response in terms of linear displacement, rotation, twisting, expansion, contraction, or change in Young's modulus (stiffness). The first step in the selection of actuators is to determine the desired performance parameters and ranges of operation. Input to an actuator can be in terms of heat, light, electrical signal, chemical reaction, electromagnetic field, pressure, physical motion, water, or any other stimulus. The general performance parameters associated with an actuator include sensitivity, resolution, stability and strength, type of output, torque, speed, operating temperature, hysteresis, transient response, and lifetime. The general parameters are used to shortlist suitable actuators to a bunch of devices from hundreds of available options in the market. Following examples will further clarify the process of shortlisting in various scenarios.

Example 5.2.3

An actuator is to be installed in a simple household water heater thermostat to maintain the temperature within the desired specified limit. The temperature ranges from −10°C to 70°C with the desired accuracy of ±5°C. Find the suitable actuators to control the gas input to fulfil the mentioned requirements.

Solution

The given scenario leaves many options open for the design engineer that are to be determined first based on the target application. It is stated that the end application is a household heater thermostat, and the gas pressure has to be controlled using an actuator to maintain a certain temperature. There are two options, first, to use a separate temperature sensor and use its data as feedback to control an independent actuator for the gas valve as is done in a modern household water geyser; second, to use an actuator directly operating on temperature as its input. Both options will be discussed one by one as possible solutions.

In the first case, modern water heaters use an electronic temperature sensor to measure water temperature, feed it to a microcontroller, and use the data as feedback to control an electromechanical actuator to operate the gas valve. Gas valves are usually controlled through a rotational motion, so, a controlled rotational actuator with electrical signal as its input is required. DC motors are a feasible option in this case. As the accuracy and holding state are important in this application, stepper motors are a better option. Linear actuators and all other forms of electromechanical actuators can thus be omitted which leaves a narrowed down list of stepper motors only that have to be checked for final selection.

The second option is still used in many old-style heaters where a bimetallic strip-based thermostat is used. Bimetallic strips are actuators that bend when heat is directly applied as the input. The changing bend angle can be used to control the gas input using a mechanical system. There are other options of thermomechanical actuators like linear shape memory alloys and shape memory polymers that convert heat directly into linear motion. Linear actuators are not usually used for this specific application.

Using a thermal actuator directly instead of an electromechanical actuators seems promising at first because no additional electronic circuitry is needed but if we further consider, additional mechanical interface is required for that and it is also not as much accurate as a modern but cheap electronic system can be.

Example 5.2.4

A car lifting mechanism is to be installed at a car wash/repair pit. The vehicles can range from small cars to SUVs. Explain the process of shortlisting a suitable actuator for that purpose.

Solution

The given example is a scenario-based problem where the design engineer has to do some research to find the desired ranges of actuator parameters. For the car lift systems, the most crucial factors are the car size and weight. The weight ranges from 1000 kg for compact cars to 3000 kg for large SUVs. The selected actuator must have the capacity to lift minimum 3000 kg weight in linear upward motion and should also hold to the final position with minimum or no energy supplied. This will straight-forward omit the possibility to use electromagnetic, chemical, thermal, and some other types of actuators. The possible candidates are high power purely mechanical geared actuators, electromechanical actuators (linear motors), and hydraulic or pneumatic actuators. Purely mechanical actuators are usually used in car jacks that are widely used to repair flat tires by slightly lifting a portion of the vehicle using a rotational or a to-and-fro motion. They are not ideal to lift a whole vehicle to a height of several feet, so, they can be omitted as well. Let both electromechanical and hydraulic or pneumatic actuators be the shortlisted candidates.

5.3 FINAL SELECTION OF DEVICES

This section will focus on how to finally select a single device out of the bunch of the shortlisted devices in the previous section that is most suitable for the application. The major things that must be considered for final selection of devices include the types of signals involved, size, cost, power requirements, and interface and conditioning circuitry.

5.3.1 Sensors

After shortlisting a bunch of sensors based on the required performance parameters and ranges, the system engineer is required to determine the types of input and output signals in the system and the working principle and type of output signal by the sensor. The signals that can be involved in a purely mechanical system include displacement, pressure, rotation, etc. In electrical system, the signals can be in the form of current or voltage. In hybrid systems, the signals can be based on pneumatic pressure, velocity, color, etc. A sensor should be able to be seamlessly integrated with the system and the output signal of the sensor should be easily readable and interpretable by the system. Also, the cost and size of the sensing device should be considered, and it should be physically and economically feasible to install the sensor in a particular system for a specified application.

Example 5.3.1

As stated in Example 5.2.1, a temperature sensor is required to be installed inside the boiler of a steam turbine. The temperature range that the sensor needs to detect is between 70°C and 400°C. The temperature needs to be detected with an accuracy of ±2°C while the transient response time should be less than 5 seconds. The lifespan of the sensor should be no less than 2 years with a maximum of 2% degradation in response during this period. Which exact sensor should be used from the shortlisted devices and how to determine that?

Solution

Based on the required performance parameters, two types of temperature sensors were shortlisted, thermocouples, and RTDs. Thermocouples provide the output in terms of an output voltage that is an electrical signal, proportional to the temperature being sensed. RTDs provide output in terms of change in electrical resistance with change in surrounding temperature. Both the feasible sensor types provide output in terms of electrical signals. Thermocouples are active devices producing their own voltage output while RTDs are passive devices that need to be supplied with voltage to measure the change in resistance that is proportional to change in temperature. In the given system, both types are OK as RTDs also consume very little power to operate that is negligible for a steam turbine. Thermocouples either require a fixed reference node or require a compensating electronic circuit to operate while RTDs have no such requirements. Size of both sensors is comparable while the cost of RTDs is higher than thermocouples. Both these factors are not important for a steam turbine system. The parameters thus left to select the suitable device are long-term stability that is excellent for RTDs; accuracy, that is again better for RTDs; output response linearity, that is again excellent for RTDs; susceptibility to noise, that is very low (better) in case of RTDs; and finally, the response time, that is much better in case of thermocouples when compared to RTDs. This means that RTD seem to be the winner here after making sure if the response time meets the minimum requirements. RTDs have response times varying from 1 to 50 seconds and the system requirements are less than 5 seconds. So, those RTDs can be used whose response time is less than 5 seconds. All what is left for the design engineer to do is to filter the RTDs with response time lower than 5 seconds and then sort them based on cost to select the final sensor.

Thermocouples could have been used in this application, but the clear better option was RTD. Thermocouples are preferable in the systems where a fast response time and low cost are crucial while the temperature ranges are much higher (usually more than 600°C).

Example 5.3.2

As stated in Example 5.2.2, a humidity sensor is to be selected for a commercial weather monitoring system where the hundreds of sensor nodes are to be spread across the city to provide information in real time. How will the final selection of sensor will be done?

Solution

The required performance parameters of humidity sensors for a weather monitoring system were discussed in Example 5.2.2. It was also discussed that there are hundreds of devices that fulfil the determined criteria. The shortlisting was based on the basic requirements and the devices either not fulfilling them or exceeding the requirements by a long shot were omitted. The data set was reduced, and the final selection was remaining. It was also discussed that the user may sort the shortlisted devices based on the cost but that was determined not to be the best option. Now, it is needed to discuss further requirements for a weather monitoring system comprising of hundreds of nodes spread across the city. First, the cost should be low as the sensor nodes will be mass deployed. Second, the size should be small, and the power requirements should be low. Third, as the sensor nodes operate on electrical signals, the sensor used should have an output compatible to be read by a simple and cheap electrical system/circuit. The required circuit should also be power-efficient and cost-effective. Based on these additional requirements, the types of humidity sensors will be examined, and the best type will be selected first. Humidity sensors with the desired performance parameters can be found to have outputs in terms of change in voltage, resistance, capacitance, surface acoustic waves, optical refractive index, color, and physical mechanical motion of diaphragm. Only resistive, capacitive, and voltaic sensors give output in terms of electrical signals, so the rest are automatically omitted. Voltaic sensors are very expensive, larger in size, and require regular maintenance but are very power efficient with simple interfacing circuit required. They are the best option if the company's financial plans and infrastructure allow, but, that is a rare case. In general, the sensors need to be small, cheap, and maintenance-free. So, only capacitive and resistive sensors are left. Both are excellent candidates for the task. The difference lies in the requirements of read-out and signal-conditioning and interfacing circuitry needed. For capacitive sensors, additional circuitry is required for signal conditioning to reliably convert the capacitance into relative humidity when compared to the resistive sensors. The circuitry is not much complicated, costly, or requires very high power. Both options have been kept open here. One cheapest sensor from both resistive and capacitive categories will be selected, and the required signal conditioning and interfacing circuits for both will be developed in later sections.

5.3.2 Actuators

After shortlisting of the actuators in the previous section, the system designed is to finally select one most suitable device for the job. The types of input and output signals in the systems need to be considered and decide which device would be the easiest to interface and will be most efficient in terms of performance, cost, and complexity. Continuation of the examples from previous section will be used to explain the process in a scenario-based problem.

Example 5.3.3

As stated in Example 5.2.3, an actuator is to be installed in a simple household water heater thermostat to maintain the temperature within the desired specified limit. Select the most suitable device from the shortlisted ones.

Solution

In Example 5.2.3, thermomechanical actuator and electromechanical actuator coupled with an electronic temperature sensor were shortlisted as the feasible options. Both these options will be kept open for the final selected devices but with specified parameters. In case of electromechanical actuators, the target is to control the position or angle of the gas valve. So, the actuator needs to have a very good position control and that can be achieved using a stepper motor instead of a simple DC motor. Secondly, the stepper motors consume zero energy while remaining at a hold position and have a good holding torque that is ideal to control the gas valve position. Accuracy can be determined by checking the angle per step for which the most common values are 3.6° and 1.8°. Accuracy in temperature control can be achieved using this system in up to fractions of a degree. So, a small stepper motor is required with enough torque to operate a mechanical gas valve and accuracy of about 1.8° per step.

In case of the thermomechanical actuator for a purely mechanical system, a bimetallic strip coupled with a mechanical assembly can be used. The bimetallic thermal actuator must have capacity to physically respond to a change of at least 5°C temperature and produce a physical movement that can be used to turn the valve on or off. A brief comparison of the features of both the devices is presented in Table 5.2.

The final selection will be based on the infrastructure available to the manufacturer and the demand of customer. In many markets, purely mechanical systems are preferred while in relatively modern applications, electromechanical systems are preferred as they allow automation and addition of many different features. Both will be discussed in the design example in the next section.

Example 5.3.4

As stated in Example 5.2.4, a car-lifting mechanism is to be installed at a car wash/repair pit with given specifications. Select the most suitable device for the task from the shortlisted candidates.

Solution

To select the most suitable device for this problem, let us first compare the available options using Table 5.3.

TABLE 5.2
Comparison of Various Parameters of Actuators for Example 5.3.3

Parameter	Bimetallic Strip	Stepper Motor
Design	Purely Mechanical	Electromechanical
Initial cost	Low	Medium
Accuracy	~ ±5°C	> ±1°C
Operation cost	Medium	Low
Expandability	Minimum	Very flexible
Efficiency	Low	High

TABLE 5.3
Comparison of Different Available Options of Actuators for Example 5.3.4

Parameter/Type	Hydraulic	Pneumatic	Electromechanical
Signal involved	Liquid	Compressed gas	Electrical
Transient response	Fast	Medium	Fast
Installation	Complex	Medium	Simple
Maintenance	High	Relatively low	Low
Operational cost	Low	Low	Medium
Weight limit	High	Medium	On lower side
Installation cost	High	Low	Medium
Required space	Large	Medium	Small

As it can be observed from the above table, an electromechanical linear actuator-based system is the most feasible option for a small-to-medium scale car wash/repair center. The required space and initial cost are the main factors here that make it difficult to install hydraulic or pneumatic actuator-based systems in these scenarios. Also, the weight limit defined in Example 5.2.4 can be easily managed by electromechanical systems. Hydraulic and pneumatic actuators are preferable for bigger stations targeting vehicles with weight ranging much higher than the defined.

5.4 SIGNAL CONDITIONING AND INTERFACING

This is one of the most crucial parts after selection of the best device suited for the system. If the best device is selected but the signal conditioning and interfacing are not done properly, the device will not be able to give the desired results and will fall short of the required performance parameters that are crucial for proper working of any type of a system. In this section, it will be discussed which parameters have to be kept in mind while designing the signal conditioning and interface circuits for sensors, transducers, and actuators. The focus will be on electrical/electronic systems considering the modern mechatronics systems almost all of which are equipped with some kind of processing/computing device (microcontroller/microprocessor).

5.4.1 Sensors

Sensors respond to the external stimuli and generate a signal of their own that is proportional to the magnitude of the stimulus being measured. The response can be in terms of change in shape, size, color, resistance, capacitance, voltage, chemical or physical properties like dielectric constant, etc. It is important to select a sensor for a system that has an output compatible with the other properties and signals involved in the rest of the system. Many a times, the sensor output is not compatible with the system as it is and needs intermediate circuitry for interfacing and signal conditioning to make it compatible with and comprehendible by the system. In this section, it will be discussed through examples, the options available for developing the signal conditioning and interface circuitry for seamless integration of sensors with the systems without compromising the required performance.

Example 5.4.1

As stated in Examples 5.2.1 and 5.3.1, a temperature sensor is required to be installed inside the boiler of a steam turbine. An RTD was selected as the most suitable device for this system based on our previous solutions. Design a signal conditioning circuit for interfacing this sensor with the system and get the final temperature as an output. The circuit should not compromise/deteriorate any of the required parameters mentioned in the example, should display the temperature for a human with a resolution of 1°C, and should be able to provide an electrical output as a feedback to control the heater intensity.

Solution

Let us assume that the selected RTD has $\alpha_o = 0.005/°\text{C}$ and $R = 500\ \Omega$ at 20°C. It is known that for RTD, the relation of output resistance with temperature (T) can be found using:

$$R(T) = R(T_o)[1 + \alpha_o \Delta T) \quad (5.1)$$

Using Equation 5.1, the range of resistance of the sensor can be found for the desired range of temperature. This gives $R(70) = 625\ \Omega$ and $R(400) = 1450\ \Omega$. Now, when one has the range of resistance for the given temperature range, the next step is to determine the range of voltage output to which the resistance must be converted/translated. Most of the commonly used industrial systems are based on programmable logic controllers (PLCs) that deal with voltage ranges of 0–24 V_{DC}. But the analog inputs of these systems take voltage input in the range of 0–V_{DC}. The first step is to convert the range of resistance to 0–5 V_{DC}. For that purpose, a bridge circuit can be used that is balanced at null for 625 Ω RTD resistance. The second step will be to find the corresponding voltage output for the RTD resistance of 1450 Ω. The third step will be to convert this voltage to 5 V_{DC}. Then, a suitable analog to digital converter (ADC) will be selected that can convert the analog voltage to digital value without compromising the resolution and response time. Finally, the converted value of temperature will be sent to a digital display. A potential divider circuit based on a DC resistive bridge can be used to perform that task. Assuming that the bridge power supply is of 5 V_{DC}, the bridge output voltage can be calculated using the following equation:

$$V_{out} = \left[\frac{R_3}{(R_1 + R_3)} - \frac{\text{RTD}}{(\text{RTD} + R_2)} \right] \times 5 \quad (5.2)$$

To null the bridge at RTD = 625 Ω, let us take $R_1 = R_2 = R_3 = 625\ \Omega$. For temperature of 400°C and RTD's resistance equal to 1450 Ω, V_{out} will be equal to negative 0.994 V. Thus, the output voltage of the bridge circuit ranges from 0 to −0.994 V that must be converted to 0–5 V to be compatible to be used as an input to the ADC. For the conversion, one needs to give a gain of negative 5.03 to the output of the bridge circuit that can be done by adjusting the variable gain resistor R_G of the instrumentation amplifier. Once the temperature has been converted to voltage in range of 0–5 V_{DC}, a suitable ADC must be selected that can convert this voltage into digital value with a resolution of at least 2°C. For $\Delta T = 2°\text{C}$, the change in

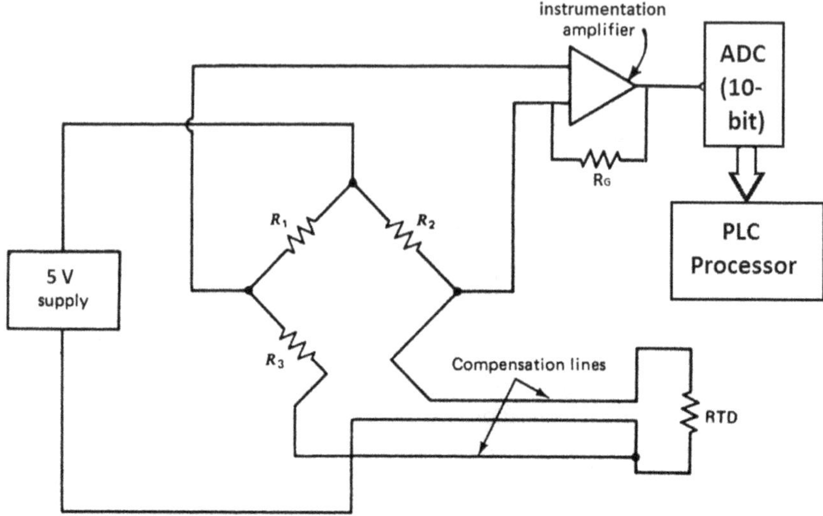

FIGURE 5.2 Final signal conditioning and interfacing circuit for Example 5.3.1.

resistance of RTD will be equal to 5 Ω. This will in return cause a voltage change of 3.994 mV at the bridge output, and 20 mV at the output of the instrumentation amplifier. The selected ADC must have a minimum resolution of 20 mV to record temperature with an accuracy of ±2°C. For a 0–5 V ADC, this range must be divided into 5/0.02 = 250 minimum steps to achieve this resolution. This means that a minimum 8-bit ADC with 256 steps can be used for this application. It is however advisable to use an ADC with a better resolution, like a 10-bit ADC that will result in a resolution of 4.9 mV or 0.5°C. Second, the ADC must have a conversion time of less than 5 seconds that is normally the case assuming that the rate of change of temperature is not very high. The complete signal conditioning and interface circuit are presented in Figure 5.2.

Finally, it was required for the circuit to provide a voltage output that can be used to control the heater. That is usually managed using the in-built features of pulse-width modulation in most of PLCs and microcontrollers/processors.

Example 5.4.2

As stated in Examples 5.2.2 and 5.3.2, a humidity sensor is to be selected for a commercial weather monitoring system where the hundreds of sensor nodes are to be spread across the city to provide information in real time. Two candidate devices, one resistive and one capacitive, were considered equally good for the said application. Describe in detail the design processes of the signal conditioning and interfacing circuits for both.

Solution

Let us start with a resistive humidity sensor with the output response curve presented in Figure 5.3a. It can be observed that the resistance of the device varies almost linearly with change in percentage relative humidity. As it is a resistive

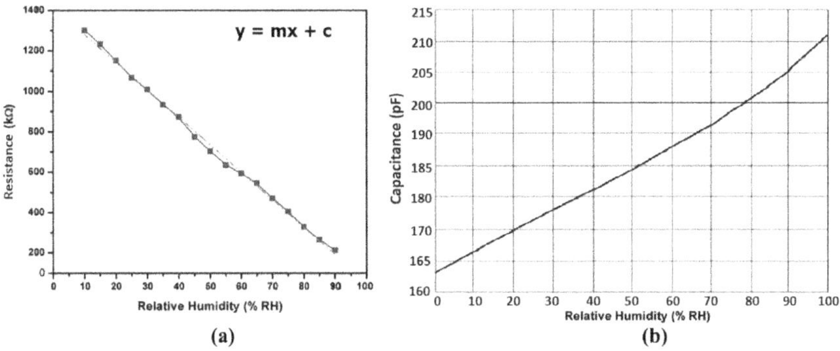

FIGURE 5.3 Output response curves of two types of relative humidity sensors: (a) resistive and (b) capacitive (HS-1101).

sensor, same process as discussed in previous Example 5.4.1 will be used to develop a signal-conditioning circuit for this sensor. The values of resistors and gain will be adjusted. Only additional point of consideration will be the power consumption that will be limited by selecting appropriate resistor values in the bridge circuit to keep the current flow within a certain allowed range.

The second humidity sensor shortlisted for this application was a commercial capacitive humidity sensor (HS-1101) with output response curve presented in Figure 5.3b. The engineer is required to design a circuit that can convert this change in capacitance into an actual value of relative humidity that can be logged into the weather monitoring system. First, the capacitance should be converted into a signal that can be taken as an input by the microcontroller without the need of any major additional components. Microcontrollers can easily read two types of signals, voltage magnitude through ADC and frequency of an AC signal using timers. Converting a change in capacitance into change in voltage with the required accuracy in this problem will require a sophisticated circuitry that is not a suitable option for hundreds of sensor nodes. Capacitance can be directly measured by charging and discharging it through a known fixed resistor using pulses and measuring the time constant RC. Another feasible and simpler approach is to use a simple square wave generator like a 555-timer circuit with output frequency dependent on the magnitude of capacitance. The frequency can be converted into capacitance using the given relationship that can be in return converted into the magnitude of %RH. The circuit diagram presented below in Figure 5.4 is one feasible solution for this problem.

In the circuit presented above, frequency is related to capacitance through a fixed formula. Capacitance can be calculated using the following nonlinear relationship:

$$C = \frac{1.44}{(R_1 + 2R_2)f} \tag{5.3}$$

And then using the linear relationship linking change in capacitance to %RH, the following equation can be used to find the actual magnitude of % relative humidity.

$$\%\text{RH} = {(C - \text{intercept})}\big/{\text{Slope}} \tag{5.4}$$

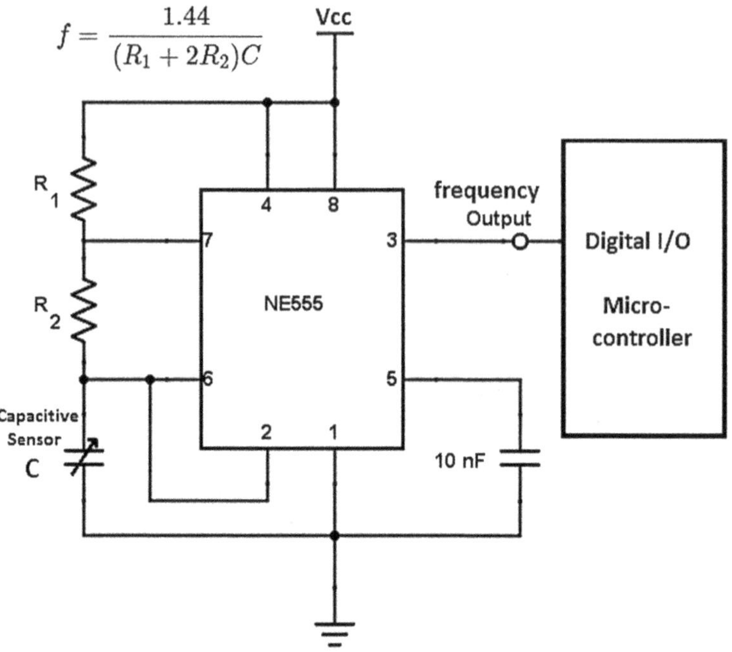

FIGURE 5.4 One possible solution for problem in Example 5.4.3.

Now, keeping in mind the requirements of resolution and accuracy, the data in Figure 4.2b will be used to find the required values of components used in our design. The capacitance changes from ~165 pF to 205 pF for 0% RH to 100% RH. This means that the slope is 0.4 pF/%RH while the intercept is 165. Our main concern is the slope as it will eventually determine the resolution. A circuit is needed that can detect a change of 0.4 pF in capacitance and convert it into a change in frequency that can be detected by the controller reliably. If $R_1 = R_2 = 30\ \text{k}\Omega$ is selected, the frequency will change in range from ~96,969 Hz to 78,049 Hz with a difference of ~18,920 Hz for the complete range of relative humidity (0%–100% RH) and ~189 Hz/%RH. The microcontroller must be able to detect a change in frequency of ~180 Hz and even the most basic microcontrollers these days can easily and reliably detect frequency changes of up to a few Hertz. This means that the designed system can be reliably used for the desired application.

5.4.2 Actuators

Actuators are somehow opposite to sensors (coupled with transducers) in a way that they provide a physical output in response to a specific input signal. The input signal can be in the form of temperature, pressure, force, mechanical motion (linear and rotational), current, voltage, light, chemical reaction, etc. It is crucial to select an actuator for a system that has its input compatible with the rest of the signals used in the system. There may be a need to develop additional interfacing and conditioning modules for seamless integration of actuators with the rest of the system. Following examples will elaborate the type of interfaces that may be required for certain real-life systems.

Example 5.4.3

As stated in Examples 5.2.3 and 5.3.3, an actuator is to be installed in a simple household water heater thermostat to maintain the temperature within the desired specified limit. Two possible candidates including a stepper motor and a bimetallic strip were selected as the most suitable devices for this system based on our previous solutions. Design the interface for both.

Solution

Two candidates were finally selected for the described application, bimetallic strip-based thermo-mechanical actuator, and stepper motor coupled with additional temperature sensor. Let us discuss the interface requirements of each of them one by one.

Starting off with the electromechanical actuator, that is a stepper motor in this case, an additional temperature sensor interface is required that can provide a feedback to the actuator based on which the actuator should control the position of the gas valve. And RTD-based sensor interface can be used for this application. The signal conditioning and interface circuit for RTD-based temperature sensor has already been discussed in Example 5.4.1 and the diagram is presented in Figure 5.2. The additional interface required for the operation of system required for this particular application is the control algorithm implemented inside the microprocessor, a stepper motor driver circuit to power the actuator that is connected to the mechanical gas valve. The position of stepper motor will control the amount of gas supplied to the burner that is heating the water and the temperature feedback of water will be fed to the microprocessor using a signal conditioning and interface circuit based on RTD. The complete system block diagram for this solution is provided in Figure 5.5

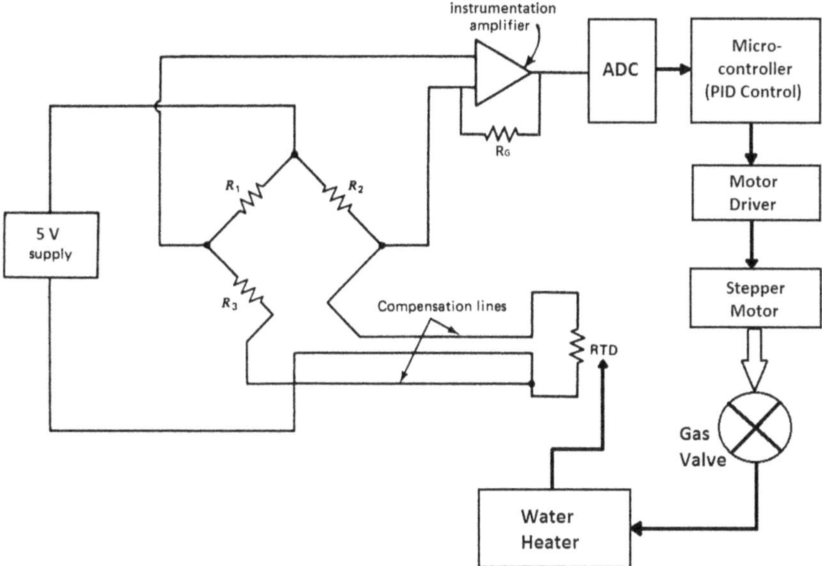

FIGURE 5.5 Interfacing and signal conditioning module with system block diagram for the first design of Example 5.4.3.

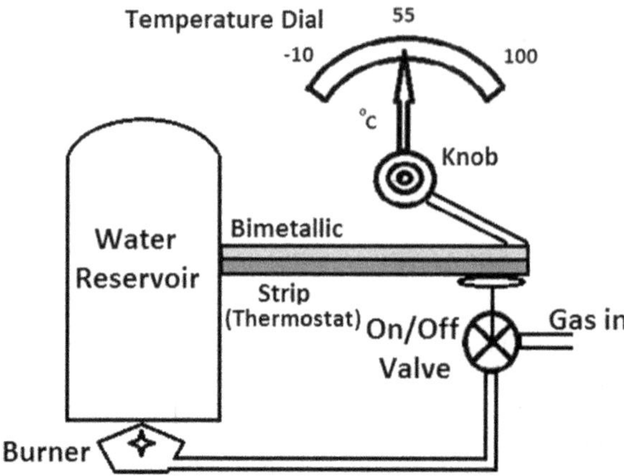

FIGURE 5.6 System block diagram for the second design of Example 5.4.3.

The second option was to use a purely mechanical system based on the bimetallic strip as the thermomechanical actuator. This actuator directly responds to the change in temperature and does not require a separate temperature sensor, conditioning/interfacing circuit, or any electrical component. The physical motion of the bimetallic switch is calibrated and is directly mechanically coupled with an on/off gas valve. If the temperature rises above the fixed value, the strip bends and the valve is turned off just like in a pressing iron. If the temperature falls below the lower limit of the set desired range, the strip will straighten and physically push the mechanical switch of the gas valve to turn on the burner and start heating again. The complete system block diagram is presented in Figure 5.6.

One thing to consider here in the second design is the natural hysteresis effect of the bimetallic strip that on one hand helps the system to be mechanically stable and avoid vibrations and malfunction while on the other hand reduces the accuracy and deals in a range of temperature rather than a desired fixed value. This makes the electromechanical actuators most sought after option in modern systems where accuracy and efficiency are much important than initial cost and manufacturing complexity.

Example 5.4.4

As stated in Examples 5.2.4 and 5.3.4, an actuator-based car lifting mechanism is to be installed at a car wash/repair pit. A linear electromechanical actuator-based system was selected for this system based on our previous solutions. Design the interface module.

Solution

In example 5.3.4, linear electromechanical actuator-based system was selected as the most suitable option for this system. To design the block diagram and interface, both electrical and mechanical interfaces are required in the design.

To implement the final design, first consider the upper weight limit in the example. To lift a 3000 kg vehicle, at least 4 hp three-phase induction motor is required. Three-phase induction motor is easier to maintain, robust in

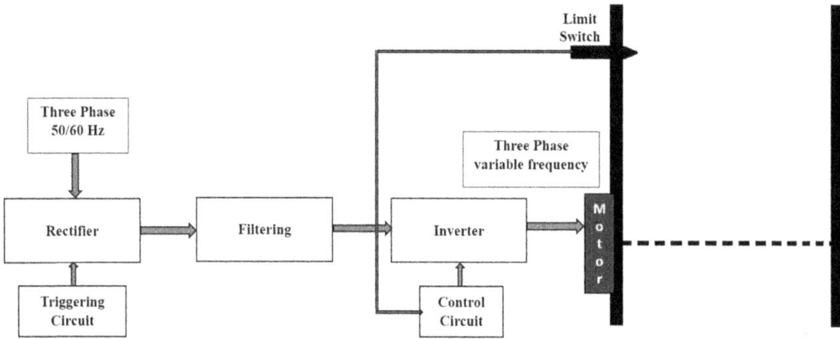

FIGURE 5.7 Interface block diagram of the linear electromechanical actuator based system for Example 5.4.4.

construction, inexpensive, self-starting, and highly reliable. To control the speed of lifter, ultimately, the speed of motor is controlled, for which a variable frequency input is required. A variable frequency drive (VFD) circuit is used to control the synchronous speed of the three-phase induction motor that can be explained by the following relation:

$$N_s = 120 \times f/P \tag{5.5}$$

where P is the number of stator poles, f is the supply frequency, and N_s is the synchronous speed of the rotor/shaft. Another method that can be used to control the speed is by changing the voltage magnitude. The block diagram of the system interface is presented below in Figure 5.7.

In the above VFD-based control circuit, the first part is a three-phase AC input that is rectified using a three-phase bridge circuit depending on the input of the triggering circuit. To get the pure DC, an additional filter circuit is added. The inverter circuit generates the desired frequency to control the speed of motor. The controller circuit at the bottom is controlling the desired frequency of the inverter and is also getting upper height feedback from the limit switch.

The above examples state one possible solution to approach the problems efficiently and simplistically at hand. There are other methods that could have been employed in the stated scenarios and could be even more efficient, but the scope of this study is to develop a feasible strategy to approach such problems.

5.5 CONCLUSIONS

The chapter clearly gives an idea of the step-by-step process that can be followed for the selection of sensors (coupled with transducers) and actuators for a certain system or application. Similar steps can be followed to select the most suitable devices for the targeted applications. The selection criteria are dependent on the performance requirements of the overall system and the parameters desired from the device. The selection can be done in two steps, shortlisting and final selection. Shortlisting is normally done based on size, cost, accuracy, and range of operation. Final selection is done based on the types on input and output signals involved in the system and the optimum results in terms of design and deployment feasibility. The selected devices are then interfaced with the actual physical system through specifically designed signal conditioning and interfacing circuits/modules. The aim in design of these modules is to minimize complexity without compromising on the performance characteristics. As a summary, no hard and fast rules exist for the selection of sensors and actuators and the process is mostly dependent on the required parameters and the approach of the design engineer, yet, a generalized flow of approaching this problem has been described in this chapter.

REFERENCES

1. M. Sajid, H. B. Kim, G. U. Siddiqui, K. H. Na and K.-H. Choi, *Sensors Actuators A Phys.*, 2017, **262**, 68–77.
2. H. B. Kim, M. Sajid, K. T. Kim, K. H. Na and K. H. Choi, *Sensors Actuators B Chem.*, 2017, **252**, 725–734.
3. G. Uddin, M. Sajid, J. Ali, S. Wan, Y. Hoi and K. Hyun, *Sensors Actuators, B Chem.*, 2018, **266**, 1–10.
4. H. M. Zeeshan Yousaf, S. W. Kim, G. Hassan, K. Karimov, K. H. Choi and M. Sajid, *Sensors Actuators, B Chem.*, 2020, **308**, 127680.
5. M. Sajid, G. U. Siddiqui, S. W. Kim, K. H. Na, Y. S. Choi and K. H. Choi, *Sensors Actuators A Phys.*, 2017, **265**, 102–110.
6. K. S. Karimov, M. Saleem, N. Ahmed, M. M. Tahir, M. S. Zahid, M. Sajid and M. M. Bashir, *Proc. Rom. Acad. Ser. A - Math. Phys. Tech. Sci. Inf. Sci.*, 2016, **17**, 84–89.
7. M. Sajid, H. B. Kim, J. H. Lim and K. H. Choi, *J. Mater. Chem. C*, 2018, **6**, 1421–1432.
8. K. H. Choi, M. Sajid, S. Aziz and B.-S. Yang, *Sensors Actuators A Phys.*, 2015, **228**, 40–49.
9. M. Sajid, S. Aziz, G. B. Kim, S. W. Kim, J. Jo and K. H. Choi, *Sci. Rep.*, 2016, **6**, 30065.
10. M. Sajid, S. W. Kim, H. B. Kim and K. H. Choi, 9th International Conference on Mechanical and Aerospace Engineering, 2018, 264–268.
11. M. Ali, I. Shah, S. W. Kim, M. Sajid, J. H. Lim and K. H. Choi, *Sensors Actuators, A Phys.*, 2018, **283**, 282–290.
12. M. Sajid, A. Osman, G. U. Siddiqui, H. B. Kim, S. W. Kim, J. B. Ko, Y. K. Lim and K. H. Choi, *Sci. Rep.*, DOI:10.1038/s41598-017-06265-1.
13 M. Sajid, J. Z. Gul, S. W. Kim, H. B. Kim, K. H. Na and K. H. Choi, *3D Print. Addit. Manuf.*, 2018, **5**, 160–169.
14. M. Mutee, M. Muqeet, R. Memoon and J. W. Lee, *J. Mater. Sci. Mater. Electron.*, 2018, **29(17)**, 14396–14405.
15. S. Wan, M. Muqeet and M. Sajid, *Thin Solid Films*, 2019, **673**, 44–51.
16. M. Sajid, H. W. Dang, K.-H. Na and K. H. Choi, *Sensors Actuators A Phys.*, 2015, **236**, 73–81.
17. Y. J. Yang, S. Aziz, S. M. Mehdi, M. Sajid, S. Jagadeesan and K. H. Choi, *J. Electron. Mater.*, 2017, **46**, 4172–4179.
18. K. H. Choi, H. B. Kim, K. Ali, M. Sajid, G. Uddin Siddiqui, D. E. Chang, H. C. Kim, J. B. Ko, H. W. Dang and Y. H. Doh, *Sci. Rep.*, 2015, **5**, 15178.
19. M. Sajid, H. B. Kim, Y. J. Yang, J. Jo and K. H. Choi, *Sensors Actuators B Chem.*, 2017, **246**, 809–818.
20. G. Hassan, M. Sajid and C. Choi, *Sci. Rep.*, 2019, **9**, 15227.
21. M. Ali, M. Sajid, M. A. U. Khalid, S. W. Kim, J. H. Lim, D. Huh and K. H. Choi, *Spectrochim. Acta - Part A Mol. Biomol. Spectrosc.*, DOI:10.1016/j.saa.2019.117610.
22. A. Arshad, K. Riaz, T. Tauqeer and M. Sajid, *2019 Int. Conf. Robot. Autom. Ind. ICRAI 2019*, 2019, 3–6.

23. J. Z. Gul, M. Sajid and K. H. Choi, *J. Mater. Chem. C*, 2019, **7**, 4692–4701.
24. R. K. Katzschmann, A. D. Marchese and D. Rus, DOI:10.1007/978-3-319-23778-7_27.
25. E. T. Roche, R. Wohlfarth, J. T. B. Overvelde, N. V. Vasilyev, F. A. Pigula, D. J. Mooney, K. Bertoldi and C. J. Walsh, *Adv. Mater.*, 2014, **26**, 1200–1206.
26. N. S. Goo, I. H. Paik, Y. C. Jung and J. W. Cho, *Smart Mater. Struct.*, 2006, **15**, 1476–1482.
27. N. S. Goo, I. H. Paik, K. J. Yoon, Y. C. Jung and J. W. Cho, *Smart Struct. Mater.*, 2004, **5390**, 194–201.
28. S. Seok, C. D. Onal, K. J. Cho, R. J. Wood, D. Rus and S. Kim, *IEEE/ASME Trans. Mechatronics*, 2013, **18**, 1485–1497.
29. S. H. Ahn, K. T. Lee, H. J. Kim, R. Wu, J. S. Kim and S. H. Song, *Int. J. Precis. Eng. Manuf.*, 2012, **13**, 631–634.
30. W. A. Khan, G. Abbas, K. Rahman, G. Hussain and C. A. Edwin, Eds., *Functional Reverse Engineering of Machine Tools*, Boca Raton, FL : CRC Press/Taylor & Francis Group, 2019. | Series: Computers in engineering design and manufacturing, 2019.

6 PCB Design and Fabrication

Fatima Mohsin Zakai, Muhammad Faizan, and Muhammad Faisal Khan
Hamdard University

CONTENTS

6.1 INTRODUCTION

In the early nineteenth century, scientists and engineers generally followed point to point construction to make a huge design on wooden or metal-based frames by attaching a large insulator to connect them. It was not a very dependable method which might cause reason to breakdowns of paths or short circuits [1].

In 1903, Hanson filed his patent which seems to be somehow like Printed Circuit Board (PCB) [1]. It included wires which were attached with conductive pieces. In1927, another researcher named Charles Ducas made some modification toward PCB design by using stencils to print wires and applying ink to possible conductivity [1]. After World War II, a person named Paul Eisler gave an idea and took another step toward PCB from the current labor-intensive practice of hand-soldering each wire on the board [2] and now it is termed as PCB.

It can be defined as a substrate which supports an electrical connection of various electrical/electronic components/devices using different tracks and other features etched on a laminated copper sheet. PCB comprises a copper layer on top of a dielectric. In case of double-sided PCB, another copper layer is present at the bottom of the substrate. Each copper layer can have deposition of 18, 35, or 70 μm and normally the overall thickness of PCB is 1.5 mm. The substrate gives rigidity and strength to the PCB which is necessary for the board.

In the present age, PCB plays an important role in manufacturing electronic circuits. It gives paths consisting of copper tracks for the current to flow in complex circuits, thus providing an efficient way of power transfer in an electrical or electromechanical system. A variety of software are commercially available for making circuit layouts. The layouts designed from these software are being used to etch out extra copper and get the layout printed on the board.

This chapter covers various aspects of PCB including basic concepts, types, layouts-designing software, etching techniques, and case studies. Different types of PCB are discussed in Section 6.2, while some commonly used software for circuit layouts are briefly described in Section 6.3. Two famous software of circuit layouts, i.e., ORCAD and Proteus, are discussed in detail in Section 6.4 while PCB fabrication process and miscellaneous details regarding PCB are covered in Section 6.5. Both wet and dry etching techniques are discussed in Section 6.6 and one case study for each technique is presented in Section 6.7. The conclusions of the chapter are summarized in Section 6.8.

6.2 TYPES OF PCBs

There are different types of PCBs available in the market, based on their manufacturing specifications, material types, and their usage. Some of the major types are discussed below:

6.2.1 Single-Sided PCBs

It is the most common and simplest type of PCBs. In this type, copper layer is deposited only on one side of the substrate. There is a dielectric substrate over which copper layer is present. When it is used for fabrication of circuit on this type of PCB, the components are soldered on one side (i.e., noncopper side), while on the other side, copper tracks are present as a result of imprinting of circuit layout. These types of PCBs are typically found in consumer electronic devices like cameras, printers, calculators, etc.

6.2.2 Double-Sided PCBs

In this type of PCBs, copper layers are present on both sides of the board. Circuit layout is prepared in the software in the form of top and bottom layers. These layers are later imprinted on both sides so that one can easily solder components. The complexity level of a double-sided PCB is more than a single-sided PCB. The applications of double-sided PCBs include vending machines, car dashboard, etc.

6.2.3 Multilayer PCBs

These types of PCBs are more complex as compared to double-sided PCBs. In comparison with double-sided PCBs, it has more layers. It has a thick design which is helpful to prevent electromagnetic interference. For making multilayer PCBs, some double-sided PCBs are compressed together after milling and drilling. Multilayer PCBs are widely used in satellite systems, weather analysis systems, GPS systems, etc. [3].

6.2.4 Rigid PCBs

These types can be used for strengthening the circuit, making them rigid, and preventing them from twisting. These PCBs can consist of ten or more layers. The ideal example of rigid PCB is motherboard.

6.2.5 Flexible PCBs

Such type of PCBs is usually required where there is a need to transform them into any shape as per requirement. The PCB can accommodate any design and is widely used in the automotive industry such as antilock braking systems, DASH (Desktop and Mobile architecture for system hardware) system, etc. This type is also used in computer electronics, medical technology, smart phones, etc. [4].

6.2.6 Rigid-Flex PCBs

These PCBs are a combination of flexible PCBs with several rigid layers. In some industries where rapid pressure, sudden movement, or some extreme conditions are required, rigid PCBs are not reliable over a long period, so at that time, these types of PCBs are used [4].

6.2.7 High-Frequency PCBs

These PCBs are generally used for transmitting high-frequency signals in the range of GHz. They are widely used in microwave and mobile applications along with several wireless applications of high frequency range. FR4, commonly used dielectric in PCBs, is not suitable for high-frequency PCB due to losses. Other dielectrics like Rogers series, etc., are used in this type of PCB, and therefore, it is costly as compared to its counterparts.

6.3 SOFTWARE FOR PCB LAYOUTS

Few decades back, the circuit designing along with layout making was done manually. With the advancement of technology and especially the computer industry, now there are number of software available for this purpose. Some of these software are briefly discussed below along with their comparison of various features.

6.3.1 OrCAD

It is one of the popular and highly recommended software for PCB designing. It is famous among engineers/technicians due to its reliability. It also provides a complete environment from the initial design to final schematic layout and helps in the growth of new entrepreneur business scales.

Key Features

- Wide variety in the designing of new modern circuit/schematic. It provides elastic changes in the upcoming future era.
- Advanced combination of integrated design simulation process.
- Easy to understand along with features of sensitive interfacing.
- Efficient for designing purposes. OrCAD also provides best real-time functional property application and also offers free built-in library.

6.3.2 Proteus

Software which is the most popular and easy to understand in designing the schematic diagram is Proteus. This software is widely used in the application of combinations of integrated circuits (ICs). Proteus helps in designing the circuit layout, simulates test, and builds expert PCB designing in the field of different engineering technologies like electronics and electrical engineering. It is exceptionally fast and simple in making the schematic diagram with the help of designing features. Proteus has perceptive combinational characteristics, generally outstanding a modern shaped layout auto-route designing feature. It is a complete software design tool in the modern era for engineering applications.

Key Features

- For designing a schematic, Proteus has almost 800 electronic components including microcontrollers, ICs, etc., in its built-in library.
- The concept of ARES PCB layout program combines strong integrated sets of different tools, which represents modern professional PCB design software.
- User clean alliances and strong layout designing.
- New and updated design features.

6.3.3 Multisim

National Instruments used Multisim is influential circuit design software that is being developed for industry requirements. This software is one of the best circuit design layout software. The SPICE simulation features are available to create the digital design circuits, which facilitates the users around the world. Multisim software is also suitable for designing the PCB layout-sat student level.

Key Features

- Evaluate both analog and digital electronics, as well as power electronics.
- Analysis of advanced and new parameters of different electronic/electrical components.
- Embedded integration for new technologies and easy design layout with prototype flow.
- User defined modified stencil, simple in use.
- More than 6,000 new components by leading manufacturers are available.

6.3.4 Eagle PCB

In this software, there are multiple features available for the designing of PCB layout. It is mostly used in different engineering applications especially for the appliances used in media industry. Autodesk option, given in this software, changes the design very easily which enhances user's confidence and expertise.

Key Features

- Convert the ideas into reality by compactable schematic editor.
- By using PCB design layout tools, life is perceptive to convert the digital schematic.
- The feature of "partial make libraries" is available for the users, which reduces time consumption in making design flow.
- Fast auto-routing tool for complex design layouts.
- Quick formation of new PCB layout (through modular block design tools) if schematic of old PCB layout is available.
- No need to take care of ball grid.

Table 6.1 shows a comparison chart of different characteristics of software discussed above which helps the readers in selecting PCB design software as per their requirement [5].

TABLE 6.1
Comparison of Various Features of Some Popular Software

Category/Criteria	Eagle PCB	OrCAD	Multisim	Proteus
Ideal for (hobbyists, professionals, everyone)	Hobbyists	Professionals	Professionals	Professionals
Price range	Free	$1,500 as deals with retailers	Free	Free
Learning curve	Medium	Limited	Low	Limited
Support/help/tutorials	High	Low	Low	High
User interface/navigation	Medium	No limit	High	High
Schematic Editor				
Placing and editing components	Medium	Medium	Medium	High
Placing and editing electrical objects (wire, ports, etc.)	Low	Medium	High	Medium
Placing and editing graphical objects	Low	Medium	High	Medium
Annotation	Medium	Medium	Medium	High
Schematic Library Editor				
Adding pins and component shape	Medium	Medium	Medium	Medium
Defining connection type and descriptions	Medium	Low	Medium	High
Multipart schematic component	Low	Low	Low	Medium
Integration with PCB footprint and PCB library	Low	Very Low	Low	High
Adding simulation properties	N/A	Medium	Low	Medium
PCB Library Editor				
Defining and placing pads	Low	Medium	Low	Medium
Custom/irregular shapes	Medium	Medium	Low	High
3D visual	N/A	Low	Medium	Medium
Defining other layers around footprint (silkscreen, keep out, solder mask, courtyard, etc.)	Low	High	High	High
Other Criteria				
Database management	N/A	Medium	High	Low
PCB routing	Low	Low	High	Low
Net management	Medium	High	N/A	High
3D visualization	Low	Medium	Medium	Low
Error debugging	Medium	Medium	Low	Low
RF design	Medium	Medium	Low	High
FPGA design	N/A	Low	N/A	N/A
File generation	High	Low	Low	High

One thing which is also important to note is that nowadays, there are many online websites available for designing and fabricating circuits on PCBs at a very lower cost. These include PCB Cart (www.pcbcart.com), PCBway.com, Sunstone (www.sunstone.com), etc.

6.4 DETAILED DISCUSSION OF TWO MOST USED SOFTWARE

In this section, two most used software available for PCB layout designing are discussed.

6.4.1 OrCAD PCB Design Software

6.4.1.1 Introduction

In the era of designing the PCB digital schematic design layout, OrCAD software is a complete package of software which provides multiple products of PCB design. It can analyze different applications of analog, digital, and mixed signal schematic circuits, with the help of three main features including schematic editor (Capture), simulator (PSpice), and the professional design PCB board digital layout solution. There are many versions available in ORCAD but latest version with complete package of PCB design is **OrCAD**® 17.2-2016.

6.4.1.2 Design Features

OrCAD professional design software with multiple features of different tools and suites can be acquired through different retailers; like one of the best product sellers in dealing the complete suite of OrCAD software is EMA Design Automation Inc. It deals OrCAD suite with multiple features of the combination tools of CAD/CAM applications in which they have the features of Capture the circuit, simulate with PSpice, and design the Layout. The function of these multiple applications can perform exclusively, showing different tools in one software, which permits for entomb tool communication features. The OrCAD software also includes different features of CAD/CAM software which can help in designing the applications like Allegro, GerbTool, or SPECTRA. Capture is the focus of the software and acts as the major feature of Electronic Design Automation tools. Capture has a wide range of different libraries of electronic components which are used to design the schematic circuit layout for simulation using PSpice and PCB digital design layout or both simultaneously. The integrated schematic circuit on captured area can be mapped in simulation PSpice model or any other physical PCB digital layout. PSpice is a complete suite of simulation computer aided engineering (CAE) tool which contains the physical as well as mathematical models to simulate the results, while digital layout for any schematic circuit is a part of CAD tools that converts components into electrical/electronic symbolic types of schematic diagram to show the physical image of the design. For designing a footprint and connecting each part in the simulation model, netlists are used to communicate between them. For more efficiency, a CAD tool is used in schematic layout functions as a front-end CAM tool, which helps in creating the data in coordination with other CAM tool function while making a PCB. As a combination of all the features of these three applications into one package, OrCAD is a powerful combination of tools to design accurate efficient schematic circuits, to simulate and to construct the electronics circuits. It is professional software which is a key to success in any project design and helps to understand making the digital design schematic layout to build the PCB on-board.

6.4.1.3 Design Flow of OrCAD Software

Generalized steps or procedure to create a PCB digital layout design that creates a schematic circuit design in Capture and then transforming the schematic capture circuit design into a digital board as Layout design are discussed in this section. The steps to create a PCB board using ORCAD software are as follows:

- Make a new file using Capture and choose to set up a new project as PCB design project with the help of using the PC Board wizard.
- At the schematic page editor window, design or create a digital schematic circuit using different applications of combinational circuit libraries in ORCAD capture features.
- Check the design rules before generating the PCB layout. Create a netlist by using capture features and save it as .MNL file for design layout.
- Arrange the components by clicking manually or auto-setup procedure feature. Open the .MNL file and select a PCB template as .TCH file. Also save it as a .MAX design project file.
- Import the .MNL netlist file into .MAX file. After creating the file, it designs a PCB track border for inserting the physical components position on the board.
- Routing is the main part to design a PCB circuit board: first arrange the components by using auto route for ideal condition and then, arrange the components path by selecting the components manual routes.
- At the end, run the feature of the postprocessor to create files, which is used to design a PCB board.

6.4.2 Proteus PCB Design Software

6.4.2.1 Introduction

Proteus Professional software is more efficient and compatible in PCB designing. It is the combination of two popular technical applications, i.e., ISIS (the schematic capture program) and ARES (the layout feature). The ISIS and ARES features make it more powerful and give a highly integrated development background in the field of designing the PCB, which is widely and frequently used for making the application of business and industries. The features of different tools which are available in Proteus professional software are very easy to use with the help of guidance manual. The features of these tools are valuable in the area of education and professional for designing the PCB digital layout. There are many advantages of Proteus professional including auto-routing feature. It also gives multiple features in different application areas like analog/digital converters techniques. It has extremely sensitive design rules in cooperative SPICE circuit simulator. This software also follows industry standards of CAD/CAM, integrated 3D viewer, etc. Proteus 8.6 is the familiar and free online version, available on different websites.

6.4.2.2 PCB Layout Designing

In Proteus professional, the features of complete integrated ARES PCB designing are available, required for any professionally applicable PCB digital layout. It is very easy to use and develop the PCB digital layout more accurately. Once the simulation is completed, user saves the schematic captured and selects the tools for designing PCB digital layout with the help of netlist of integrated ARES suit.

6.5 PCB FABRICATION PROCESS

A PCB board comprises of two things: first, a board known as substrate, and second, the copper layer over the substrate. The substrate supports the circuit components onto the board and the tracks of printed copper wire traces are like footprint of the components which provides electrical insulation between conductive areas of the physical components.

6.5.1 PCB Substrate/Material

PCB is the most essential and important part in the design of electronic circuits. The selection of PCB material requires attention with complete knowledge of the function of the circuit. PCB design board has four layers, which combine with heat lamination and convert into one signal layer. The multiple materials used in PCB board from upper layer to lowest layer are silkscreen, solder-mask, copper, and substrate.

There are three commonly used substrate materials used in the fabrication of PCB boards which are explained below:

6.5.1.1 FR4

Most circuit boards are made of a material known as FR4 (fire retardant). This is a common type of substrate material, which provides main solid base for designing the PCB board. FR4 is manufactured by fiberglass–epoxy laminate, which helps to make the PCB easily. FR4 has the same features of fiberglass boards except only change in due to its blaze resistant. It varies according to the application logic of circuit design. FR4 material has a base standard in the substrate. Its major advantages include reliability, cost, electronics/electrical properties, and standard of performance efficiency.

6.5.1.2 PTFE (Teflon)

Polytetrafluoroethylene (PTFE) is another type of PCB board material which is manufactured by plastic material with no effect of any kind of resistance. This type of board is designed for high-speed applications as well as for high-frequency circuits design like RF circuits, etc. PTFE is exceptionally flexible in the fabrication process. It is recommended in the application designing of industries due to its lightweight fabrication board. PTFE has versatile features in the fabrication process like strong physical strength, lower dielectric loss, flame resistance, temperature stability, low fabrication cost in circuit design, etc. Rogers is the best manufacturing company corporation which provides laminate materials PCB circuit boards. Microwave and high-frequency PCB materials are sensitive toward dielectric properties and they are the best examples of Rogers PCB materials [6].

6.5.1.3 Metal

Nowadays the common materials like copper, aluminum, iron, etc. are also used in the fabrication of PCB board design. These types of materials permit the designing of Surface Mount Technology (SMT) for fabricating circuits. These metals have mechanical strength, and the life of metal base PCB is much longer.

6.5.2 PCB Layer Stack-up

Stack-up is a process of setting the wire track route of copper layers on the PCB in the process of making layout design. By using the stack-up principles, the PCB board allows to permit multiple integrated design circuits on one (single) board through multiple PCB layers. There are several advantages of PCB stack-up designs such as:

- It minimizes the circuit with less exposure of external noise.
- Low impedance and cross-talk features in high-frequency PCB layouts circuits.
- Low cost in designing the PCB of high-speed frequency circuits.
- In communication signals, it has an efficient method for making multiple layers of PCB.
- It modifies the feasibility of electromagnetic features.

6.5.3 Multilayer PCB Fabrication Advantages

The modern era has introduced IC design applications in the field of electronic engineering. The fabrication of ICs into a signal layer is very difficult to design, so it is desirable to fabricate complex circuits in multiple layers board. The multilayer technique board solves different design issues and problems such as crosstalk, noise, capacitive coupling, etc., which increases the efficiency and resolution of ICs. The performance of these layers in designing analog/digital circuits is highly efficient. The main purpose of making multiple layers PCB board is designing of highly sensitive or complex circuits. There are some advantages to use multilayer PCBs:

- Low noise resolution
- Small as well as light in weight
- Low impedance
- Fastest speed of signal transmission
- Enhanced shielding effect
- Advanced assembly density

6.5.4 Track Distance

Design rules for intertrack distance/spacing are different for low- and high-frequency PCBs. For layouts of high-voltage circuits, intertrack distance needs special attention. Voltage arc can be produced if the voltage between the metallic tracks exceeds the breakdown voltage. To avoid this condition, there are two parameters (clearance and creepage distance) which must be taken into consideration.

The shortest distance through air between two conductors is known as clearance. It depends on the material of the PCB board, potential difference, and the atmospheric nature (environmental condition). Humidity and dust change the breakdown voltage of air, so it is necessary to avoid these factors, thus reducing the possibility of shortening of the tracks.

Creepage is another parameter which helps in measuring the distance between the two components placed on conductors of a PCB. It measures the shortest nearby distance with the surface of the insulation materials. It depends on the board material as well as the environmental condition. The creepage and the clearance have same effect of moisture which is crucial in deciding intertrack distance [7]. The IPC-2221 standard is used to calculate track width for various values of current [8].

6.6 ETCHING TECHNIQUES

Etching is the process in which a circuit diagram is engraved on the metal and conductive tracks are being made through which current can flow. Generally, there are two major types of etching: (i) wet etching and (ii) dry etching; both are discussed below:

6.6.1 Wet Etching

Wet etching process is relatively cheaper as compared to its counterpart and is commonly used. In this section, wet etching is thoroughly discussed including the steps required to perform it. Wet etching can be performed using two methods; first is screen printing method in which conductive ink is used to make tracks for flow current, and after marking, it is placed in a chemical to remove excessive copper.

Another method is iron-printed glossy paper method. In this method, after circuit designing, its layout is made using any available software like Eagle, Proteus, OrCAD, etc. In the next step, circuit layout is printed on a glossy paper through a laser printer. The print is then cut from edge to edge, placed on a single layer PCB, and is being ironed for approx. 8–10 minutes [9].

Due to ironing, the print will appear on the board. For removal of extra copper, PCB is placed into ferric chloride ($FeCl_3$) solution and is being shaken up to 10 minutes. Unwanted copper is now etched out, and as the last step, sandpaper is rubbed on the PCB for smoothening/shining and now the components can be soldered on PCB for usage. The major steps of wet etching are shown in Figure 6.1a and b [10].

6.6.2 Dry Etching

There is another way of etching which is named as dry etching technique. It is being done using a milling machine where the extra/unwanted copper is removed by using milling bits. This technique is costly as compared to wet etching, but it provides more precision. It is not being used for mass production and mostly preferred for prototype circuits [9].

There is a number of PCB milling machines available in the market and the major working steps are somehow the same. For understanding the concept of dry etching technique, it is being explained in this section using LPKF ProtoMat S series machines. A photograph of the S-100 model of this series is shown in Figure 6.2.

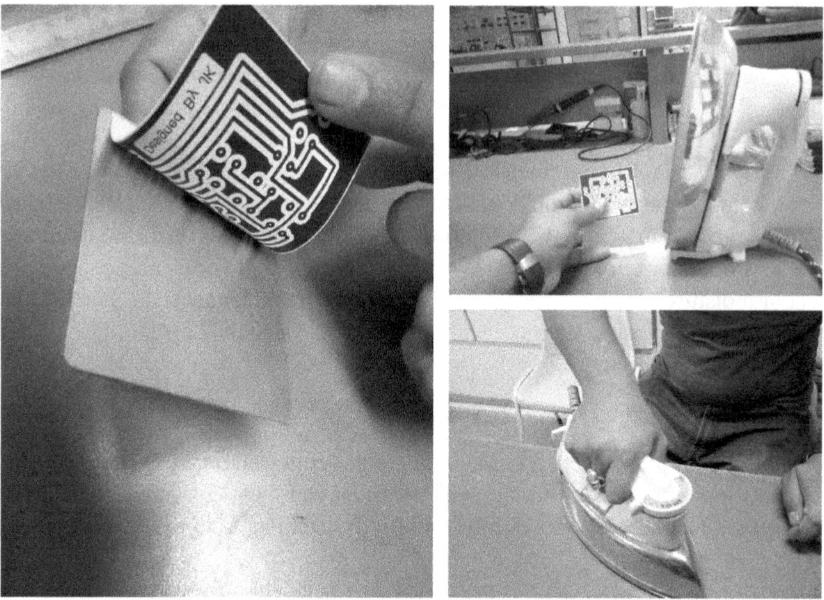

FIGURE 6.1A Initial steps of wet etching.

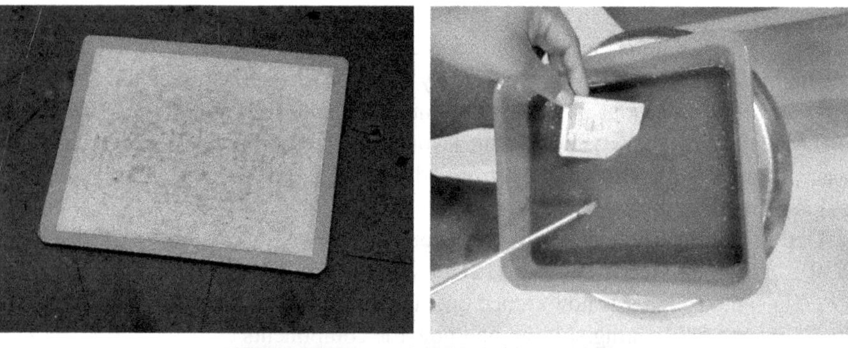

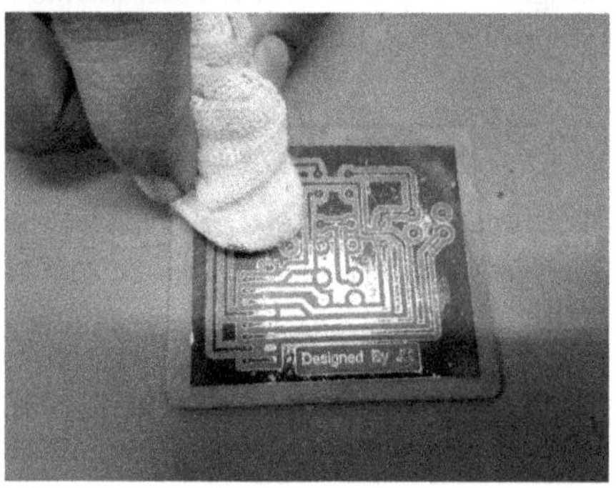

FIGURE 6.1B Final steps of wet etching.

FIGURE 6.2 LPKF ProtoMat S-100 machine. (*Photograph Courtesy*: Faculty of Electrical Engineering, Ghulam Ishaq Khan Institute of Engineering Sciences and Technology, Topi, Pakistan.)

The first step of this process is the same as that of wet etching technique, i.e., circuit designing using any CAD software (few are discussed in Section 6.3). The chosen software must have the capability of exporting the circuit layout in GerberX format. For single-sided PCB, there will be only one GerberX file for top layer, a board outline file, and a drill file which has the information regarding the number of drills along with their sizes. In case of dual-sided PCB, there will be two GerberX files (i.e., one for each layer) while remaining two files will be the same.

These files are imported in the software named CircuitCAM (provided with LPKF machine) which calculates tools/bits required for milling and drilling on PCB. The cutting of PCB is also finalized in this software along with finalization of bit that will perform this task. Later on, an export file with LMD extension is generated using CircuitCAM.

As the last step, .LMD file is being imported in software named BoardMaster (provided with LPKF machine) which is the software to run the milling machine. Before using BoardMaster, there is no need to turn on the machine, but the machine must be switched on while using BoardMaster [11]. After importing the .LMD file, the software automatically checks whether the required bits are available in the machine or not. If not, then it will give option to place the desired bit inside the machine. Before starting milling, the machine also provides provision to set the depth of the bit to make sure removal of all unwanted copper. In case of double-sided PCB, after milling, the PCB sheet has to be turned around for milling of another layer of the design. After milling and drilling both sides, the desired circuit is being cut from the complete PCB sheet for soldering of various components as per circuit requirement.

The working steps of dry etching are explained in Figure 6.3, taking LPKF ProtoMat S series machines as example. Dry etching is highly recommended for the circuits where precision is the topmost priority like RF circuits, etc., but the cost of the machine is the main hurdle in its popularity among consumers. Therefore, the user has to make an optimized decision among wet and dry etching.

6.6.3 Surface Mount Technology

While discussing about the etching techniques, there is another topic related to PCB, named as SMT, which enables to control manufacturing process easily. Its main advantage is to accommodate small components in such a way that through PCB it can be more complex for small surface area. Apart from this, SMT also allows simpler and faster automated assembly. Some plasma machines can place more than a hundred thousand components per hour for SMT.

6.7 CASE STUDIES

For better understanding of any topic, case studies are always very helpful; therefore, two case studies (one for each etching technique) are discussed in this section to get in-depth knowledge regarding PCB design and fabrication.

6.7.1 Internet of Things-Based Water Monitoring System (A Case Study)

Water is one of the basic requirements of a human being. Keeping in mind that very less part of water available on earth is drinkable, water must be properly managed and monitored. In most of the areas, there are small or large reservoirs which are many times far away from the office area, so it is difficult to monitor their storage and water quality every reservoir.

Therefore, an industrial project was done related to this task as a joint effort of Karachi based company named DATALOG and Hamdard University in 2019–2020, while the funding was provided by National Cleaner Production Center Pakistan. The solution provides a cheaper solution of the problem as compared to the SCADA system.

The main task of the project is to acquire real-time data of different parameters of water such as turbidity, total dissolved solvent (TDS), ambient temperature, and water level for monitoring at remote end. For this purpose, Internet of Things concept was used to transfer data from site to office at the far end.

A box containing sensors was deployed at the site to perform the desired operation. A circuit was designed first, incorporating all the sensors, and then a layout of the circuit was made in CAD software. Later on, using steps of wet etching, the PCB was fabricated, and after soldering of the components, it was deployed on the site for testing and debugging.

6.7.2 Ultrawideband Vivaldi Antenna (A Case Study)

In this section, fabrication of ultrawideband Vivaldi antenna is being presented as a case study. Microwave circuits always require high-precision geometric parameters; therefore, this antenna was manufactured using dry etching through LPKF ProtoMat S-100 machine.

The design of the antenna was taken from Ref. [12]. As this antenna needs metallic parts (i.e., copper) on both sides of PCB, therefore while making its layout in CAD software, two layers (top and bottom) were prepared. No drill is being required in this antenna; therefore, there was no drill file exported from CAD software. After layout designing, all three files, including gerberX files of top and bottom layers along with boardoutline file, were imported in CircuitCAM software for calculation regarding bits. An export file with extension .LMD was generated by the software and was imported in BoardMaster software. After adjusting bit depth of the machine using BoardMaster software, milling of top layer was done, and after completing it, PCB sheet was turned around and the same procedure was repeated for bottom layer. In the end, the fabricated antenna was cut from the PCB sheet and SMA connector was being soldered so that high-frequency source cable can relate to the antenna [13].

A photograph of the fabricated Vivaldi antenna is shown in Figure 6.4 in which copper at top layer and SMA connector are clearly visible, while copper of bottom layer can be observed as a shaded part. This antenna was manufactured by a research group at Ghulam Ishaq Khan Institute of Engineering Sciences and Technology Pakistan and the results using the antenna were published in Ref. [14].

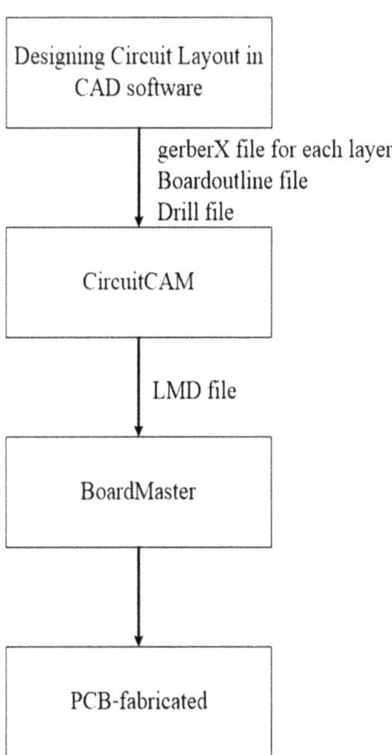

FIGURE 6.3 Flowchart explaining PCB fabrication steps using LPKF ProtoMat S series machines.

FIGURE 6.4 Photograph of fabricated Vivaldi antenna.

6.8 CONCLUSIONS

PCB provides the most reliable platform to house electrical/electronic components for a circuit. Its rigidity and compactness distinguish it from vero and bread boards. This chapter summarizes knowledge of almost all the important topics regarding PCB. Types of PCB software used for circuit layouts have thoroughly been discussed. Comparison among PCB design software broadens reader's knowledge about PCB. The procedure of both etching techniques has been thoroughly presented, highlighting every important part of it. Two case studies are also part of the chapter that clarifies the significance of the topic. This chapter provides a comprehensive knowledge to the readers about PCB and its related topics

REFERENCES

1. "The History of Printed Circuit Boards" – Infographic https://pcb-solutions.com/blog/pcb-market-monitor/the-history-of-pcb-infographic/#:~:text=The%20first%20printed%20circuit%2boards,directly%20on%20an%20insulated%20surface
2. Silvestre Bergés S, Salazar Soler J, Marzo J. Printed Circuit Board (PCB) design process and fabrication.
3. "Multilayer Printed Circuit Board- PCB" |Amitron at https://www.amitroncorp.com/printed-circuit-boards/multilayer.html
4. "PCB Materials Selection Guide", available at //www.pcbcart.com/pcb-capability/pcb-materials/ uploads/08/02/2017/, accessed date Dec 2019.
5. "The Ultimate PCB Design Software", Comparison Guide, https://www.sfcircuits.com/pcb-school/pcb-design-software-comparison-guide, accessed date Dec 2019.
6. "Printed Circuit Board PCB Materials", Twisted Traces, //www.twistedtraces.com/capabilities/pcb-materials// uploads/16/01/2020

7. "Introduction to OrCAD Capture and PSpice", http://userweb.eng.gla.ac.uk/john.davies/orcad/spiceintro160.pdf
8. Wang Y, De Haan SW, Ferreira JA. Thermal design guideline of PCB traces under DC and AC current. In *2009 IEEE Energy Conversion Congress and Exposition* 2009 Sep 20 (pp. 1240–1246). IEEE.
9. "How to Make PCB" at home available at /https://circuitdigest.com/article/how-to-make-a-pcb-at-home
10. Balasubramaniam A. "PCB Design", IEEE Seminar, available at: https://aswinbala.me/PCB_Design_WS_PPT.pdf
11. Khan MU, Sharawi MS. "PCB Fabrication Using LPFK ProMat® S62", User Manual, April 2012, King Fahd University of Petroleum and Minerals, Dhahran, KSA.
12. Mehdipour A, Mohammadpour-Aghdam K, Faraji-Dana R. Complete dispersion analysis of Vivaldi antenna for ultra wideband applications. *Progress in Electromagnetics Research.* 2007;77:85–96.
13. Khan MF, Mughal MJ, Bilal M. Effective permeability of an S-shaped resonator. *Microwave and Optical Technology Letters.* 2012 Feb;54(2):282–6.
14. Khan MF, Mughal MJ, Bilal M. Rotation—A technique to tune the working frequency of left-handed materials. *Microwave and Optical Technology Letters.* 2011 Nov;53(11):2517–21.

7 Vibration-Based Piezoelectric Energy Harvester for Wireless Sensor Node Application

Muhammad Iqbal and Malik M. Nauman
Universiti Brunei Darussalam

Farid U. Khan
University of Engineering and Technology

Asif Iqbal
Universiti Brunei Darussalam

Muhammad Aamir
Edith Cowan University

A. E. Pg. Abas and Quentin Cheok
Universiti Brunei Darussalam

CONTENTS

7.1 INTRODUCTION

The latest drive in low-power microelectronics, like wireless sensor nodes (WSNs) and mobile electronic gadgets, motivated research efforts toward motion-based kinetic energy harvesting. This is because of the increasing needs of these devices in the current era of the Internet of Things as well as the advancement in technologies, allowing the harvesting of energy from various ambient sources. Energy harvesting devices can either be used as a replacement of electrochemical batteries [1] or enhance the lifespan of these batteries, to increase the capability of a WSNW [2] in deserted and remote applications. The batteries owing to their limited lifespan need to be replaced repeatedly, which restricts the utility of WSNs in embedded and remote locations [3].

Energy harvesting is the process by which ambient energy can be converted into electricity for low-power WSNs and consumer electronics [4]. Different means of harvesting energy include thermal [5], wind [6], solar [7], acoustic [8], and nuclear reaction [9]. Table 7.1 provides information about various ambient energy sources, along with their power densities.

Generally, harvesting energy from mechanical vibrations is one of the most promising technologies [25], with ambient vibration energy obtained from typical means, such as mechanical vibrations of human motions [26], automobile and airplanes [27], household goods and machines [28], buildings and bridges [29], ocean waves [15], and other civil entities [30], is converted into electrical energy to provide a sustainable power to microelectronics.

Vibration-based energy harvesting mainly includes piezoelectric energy harvesters (PEEHs) [31–34], electromagnetic energy harvesters (EMEHs) [35–38], electrostatic energy harvesters (ESEHs) [39–42], and triboelectric energy harvesters (TEEHs) [43–46]. Figure 7.1 provides the general conversion principle of a vibration-based energy harvester. Generally, in the vibration energy harvester, ambient mechanical vibration energy is transformed into amplified periodic motion consisting of the conversion of kinetic into potential energy and vice versa, which is then converted into electrical energy with the implementation of a suitable energy transduction mechanism. In the case of PEEHs, the kinetic ambient energy puts the piezoelectric materials under stress/strain, resulting in the generation of potential

TABLE 7.1
Energy Harvesting Opportunities from Different Environmental Sources

Energy Source	Power Density	Unit	Refs.
Airflow	1	μW/cm^2	[10]
Acoustic noise	0.96 (100 dB)	μW/cm^3	[11]
Ambient sun light	100	mW/cm^2	[12]
Ambient illuminated light	100	W/cm^2	[13]
Ambient radio frequency	1	μW/cm^2	[14]
Flow-induced vibrations	15	W/cm^2	[15]
Human body movements	4	W/cm^3	[16]
Heel strike	7	W/cm^2	[17]
Handshaking harvester	30	W/kg	[18]
Insole energy harvesters	330	μW/cm^2	[19]
Machine vibration	800	W/cm^3	[20]
Vibration (piezoelectric)	200	μW/cm^3	[21]
Pushbutton	50	J/N	[22]
Thermoelectric	60	W/cm^2	[23]
Thermal	10–40	μW/cm^3	[24]

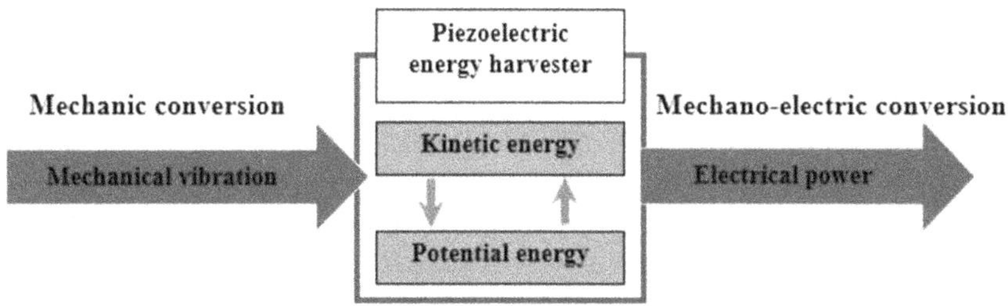

FIGURE 7.1 Schematic diagram of power flow through a vibration-based PEEH.

difference (voltage) through ordered orientation and alignment of the positive and negative charges across the material. PEEH is commonly implemented as an oscillating system, with a proof mass attached to frequency-tune the harvester and to provide the exaggerated displacement and, hence, inducing higher strain within the piezoelectric materials. The polarization due to the induced strain results in electric charge flow which depends on the vibration intensity of input kinetic energy.

The mechanical vibration energy can be converted (through mechanical mechanism) into electrical energy (through mechanical-electrical transduction mechanism) with direct piezoelectric effect [47]. PEEHs among the other motion-based energy harvesting technologies are more significant than their counterparts (EMEHs and ESEHs). These transducers are probably the oldest and attracted great research interest due to their simple design, architecture, higher electromechanical coupling, easy implementation, and vast applications in wearable, implanted, and autonomous WSNs [48].

ESEHs are usually a variable capacitor with a pair of initially charged electrodes that are used to generate charges (or enhanced the voltage) from the relative motion caused by the ambient kinetic energy source [49], while EMEHs take advantage of electromagnetic induction due to the relative motion between a magnet and a coil [50]. However, on the other hand, TEEHs combine triboelectrification and electrostatic induction for its conversion of kinetic ambient energy to electricity [51].

The comparison of vibration-based ESEHs, EMEHs, and PEEHs is listed in Table 7.2. Comparatively, due to the need of one piezoelectric layer as the transduction medium, PEEHs are simple to design in any permissible architecture. Moreover, the deposition of the piezoelectric layer by the sol-gel process makes the fabrication of PEEHs more suitable to be integrated with the standard microelectronics and MEMS technologies. In contrast to ESEHs and EMEHs, the scaling down (micro- and nanoscale development) approach works well with the fabrication of PEEHs. Like ESEHs, the pull-in phenomenon does not occur in PEEHs and these do not require an extra battery for the initial charging. ESEHs and PEEHs can produce relatively higher voltage outputs than their counterpart (EMEHs). Moreover, since the output voltage levels that are produced in EMEHs are on the lower side, as a result, the AC-DC voltage conversion is challenging, and usually, EMEHs require special ultralow voltage and ultralow power AC-DC convertors. On the contrary to EMEHs, the internal impedance of ESEHs and PEEHs is relatively large because of which small levels of current output are obtainable from the later energy harvesters.

TABLE 7.2
Vibration-Based Energy Harvesters' Comparison

Parameters	ESEHs	EMEHs	PEEHs
Architecture	Difficult	Difficult	Simple
Integration with MEMS technology	Moderate	Difficult	Simple
Integration with microelectronics technology	Moderate	Difficult	Simple
Fabrication	Difficult	Moderate	Simple
Micro- and nanoscaling	Possible	Difficult	Possible
Compatibility with integrated electronic devices	Difficult	Difficult	Simple
Initial charging	Necessary	Not required	Not required
Pull-in phenomenon	Present	Not present	Not present
Voltage generation	High level	Low level	High level
Current generation	Low level	High level	Low level
Internal impedance	Large	small	large
Natural frequency	High	Low	High
AC-DC voltage conversion	Simple	Difficult	Simple
Voltage switching circuit	Necessary	Not required	Not required
Sensitivity to vibrations	high	low	high
Power density	high	low	high

Furthermore, one of the major disadvantages of ESEHs is the need of an extra battery for the uninterrupted initial charging of the conductive plates at the beginning of the operational cycle in these harvesters; moreover, also an electrostatic-energy extorting circuit (voltage switching-circuit) is necessary to timely accumulate the transformed energy from conductive plates of the ESEH.

These harvesting mechanisms may also be used to sustainably power WSNs for nonstop condition monitoring of critical infrastructure. Base acceleration and frequency levels of common mechanical vibration sources are presented in Table 7.3.

The chapter is organized into the following sections as Section 7.2 summarizes different piezoelectric materials and their application to energy harvesting. Section 7.3 explained vibration-based PEEHs in detail. The developed and reported

TABLE 7.3
Vibrational Characteristics of Various Ambient Resources

Vibration Source	Vibration Frequency (Hz)	Base Acceleration (g)	Refs.
Human walking	0–5	0.2–5	[52]
Human motion	5–10	0.1–10	[53]
Bridges	1–40	0.1–3.79	[54]
Automobiles	15–50	0.3–2	[55]
Home appliances	51–200	0.01–10	[56]
Industries	> 200	1–10	[57]

PEEHs to date are compared in Section 7.4, and lastly, Section 7.5 concluded the chapter with future recommendation remarks.

7.2 PIEZOELECTRIC MATERIALS

Piezoelectric materials can mainly be found as crystalline structures and become electrically polarized when subjected to mechanical strain through direct piezoelectric effect. The polarization in these materials is directly proportional to the applied mechanical stress [10]. Common piezoelectric materials including the naturally occurring single crystals, such as quartz, thin-film (zinc oxide), piezoceramics (PZT), polymers and its derivatives such as polyvinylidene fluoride (PVDF), and artificial crystals, such as barium titanate, gallium orthophosphate, and ammonium dihydrogen phosphate can be used in PEEHs. Among the piezoceramics and polymers, lead zirconate titanate (PZT) and PVDF, respectively, are materials of choice for macro- and micro-scale energy harvesting systems due to their high energy conversion efficiency. Comparatively, PZT is denser (7.6 g/cm^3) than PVDF (1.77 g/cm^3) and is increasingly used in the development of small-scale and lightweight PEEHs due to its compatibility with microfabrication technology [58]. Additionally, PZT provides a good signal-to-noise ratio, and has a high energy density, large piezoelectric coefficient, and dielectric constant in a wide dynamic range than other piezoelectric counterparts, such as ZnO and AlN. Due to its comparative high stiffness, PZT is often used as a bimorph structure that favors bending, such as a cantilever beam with tip-proof mass to match the low-frequency environmental vibrations (Table 7.4).

Within the context of mechanical-electrical conversion, the main property of piezoelectric materials for usage in vibration energy harvesters is piezoelectric charge constant (d). It is the polarization generated per unit of mechanical stress applied to a piezoelectric material and is commonly expressed in Coulomb/Newton [70].

$$\mathrm{d}U = \sigma_{ij}\mathrm{d}\varepsilon_{ij} + \mathrm{E}_i\mathrm{d}\,\mathrm{D}_i \tag{7.1}$$

where $\mathrm{d}U$ is the change in energy density, σ_{ij} is the stress tensor, and ε_{ij} $(i.j = 1, 2, 3.)$ is the strain tensor. E_i and D_i are the electric field vector component and electric displacement vector component, respectively.

As the strength of polarization is dependent on the applied stress to the piezoelectric material as well as the value d; therefore, d is an important indicator of a material's suitability to function as mechanical to the electrical converter in vibration-based PEEHs

7.3 PIEZOELECTRIC ENERGY HARVESTERS

PEEHs may consist of a flexible piezoelectric membrane, and when the harvester is subjected to base excitations, the membrane starts oscillations that cause deformation in the material and voltage is induced by polarization (dipoles alignment) through direct piezoelectric phenomenon. PEEHs may also be composed of a cantilever beam with a tip-proof mass attached to the beam. The deflection of the beam due to the tip mass results in displacement amplification and the stored strain energy

TABLE 7.4
Classification of Commonly Available Piezoelectric Materials

Piezoelectric Material		IUPAC Name	Chemical Formula	Conversion Mode	Crystal Form	Piezoelectric Constant (pm/V)	Refs.
Ceramic	Lead-based ceramic	PZT	Pb {Zr.Ti}O_3	d_{33}	Polycrystalline	117	[59]
		Lead Magnesium Niobate	PZT (PMN-PZT)	d_{33}	Polycrystalline	1500	[60]
		Lead lanthanum zirconate titanate	$Pb_{0.92}La_{0.08}(Zr_{0.52}Ti_{0.48})O_3$	d_{33}	Polycrystalline	545	[61]
		Zinc oxide	ZnO	d_{33}	Single crystalline	12	[62]
	Lead-free ceramic	Quartz	SiO_2	d_{11}	Single crystalline	2.3	[63]
		Gallium orthophosphate	$GaPO_4$	d_{11}	Single crystalline	4.5	[64]
		Barium titanate	$BaTiO_3$	d_{15}	Single crystalline	587	[65]
		Ammonium dihydrogen phosphate	H_6NO_4P	d_{36}	Single crystalline	48	[66]
		Potassium niobate	$KNbO_3$	d_{33}	White rhombohedral crystals	91.7	[67]
		Bismuth Sodium Titanate	BNT-BKT	-	Single crystalline	-	[68]
Polymers		PVDF	–(C2H2F2)n–	d_{31}	Semicrystalline	28	[69]

in the beam is converted into electricity which is then passed on to the external circuit [71]. Some of the applied energy in the energy harvester is wasted as dielectric losses, air damping, and material damping. As environmental vibrations are mostly random and low frequency in nature, that is why vibration-based energy harvesters require frequency matching (under resonance condition) during operation, and these need to function at lower frequency and acceleration levels for optimum electrical output production (Figure 7.2).

To achieve lower resonant states in PEEHs, a seismic mass is typically attached at the free end or a slender bimorph beam is used. Human motion-induced vibrations are characterized by low frequency (which never exceeds few hertz) and high displacement amplitude, and the frequency mismatching is a real challenge in human-based harvesters [72]. Therefore, human-powered energy harvesters depend on the amount of available energy, harvester's efficiency, the conversion mechanism, and

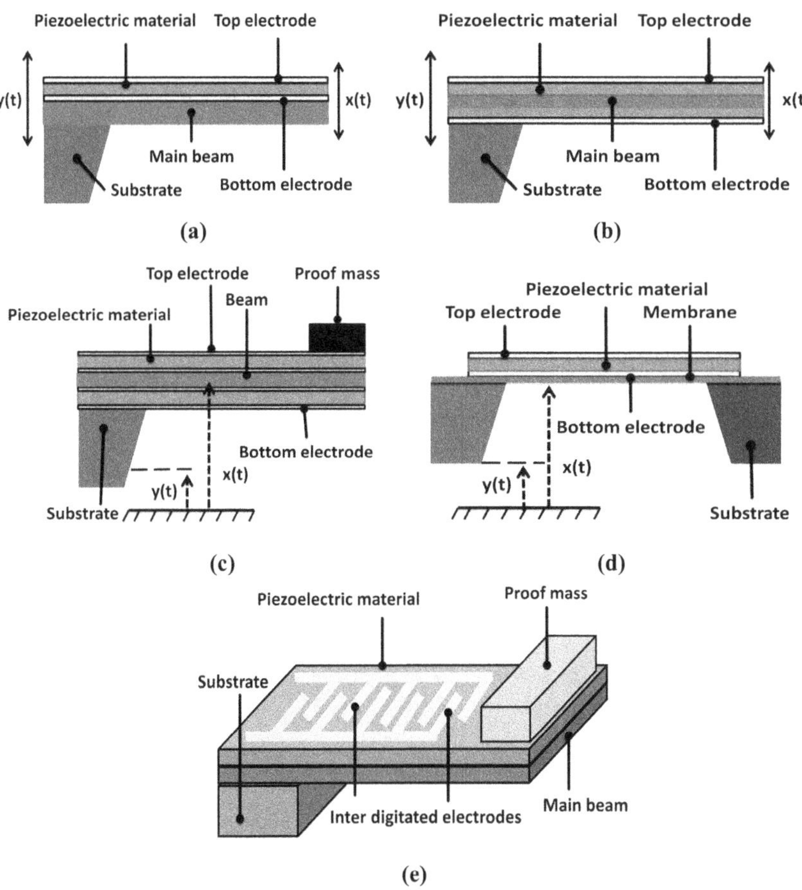

FIGURE 7.2 Architectures of resonant PEEHs: (a) unimorph with no proof mass, (b) bimorph with no proof mass, (c) bimorph with proof mass, (d) membrane type, (e) beam type with interdigitated electrodes.

the relevant electronics [73]. In the literature, mostly PEEHs are resonant type; however, few nonresonant PEEHs are also reported. Energy harvesters responding at magnified amplitudes due to matching of the natural frequency of harvester with the external excitation frequency are categorized as resonant PEEHs. Resonant PEEHs can either be linear or nonlinear. Linear resonant harvesters perform well at resonant frequencies, while these are less efficient during off-resonant conditions [74]. Nonlinear energy harvesters respond quickly to a frequency change even at low amplitude vibrations and operate more effectively in low-frequency excitations and relatively over wider frequency bandwidth. However, nonresonant PEEHs are fit to be operated at any frequency within the broad-frequency band of the harvester and need no frequency tuning and respond efficiently over a broader range of operating frequencies.

Common PEEHs typically work on two types of mechanical-electrical conversion modes, i.e., d_{31} and d_{33} modes. Of the two subscripts in the notation of these modes, the first subscript indicates the direction of polarization generated in the material while the second subscript indicates the direction of the applied stress.

In d_{31} mode, a lateral force is perpendicularly applied to the direction of polarization as shown in Figure 7.3. The applied force is however in the same direction with the direction of polarization in d_{33} mode. Practical examples of both d_{31} and d_{33} conversion modes are bending of a cantilever beam and compression of a piezoelectric block, respectively, with electrodes on its top and bottom sides in both modes [75]. Generally, d_{31} mode has a lower electromechanical coupling coefficient than d_{33} mode.

Analysis, design, and fabrication of two MEMS-scale PEEHs with top and bottom electrodes are reported in reference [76]. In the harvesters PZT and silicon dioxide (SiO_2) layers were deposited on SiO_2 substrate using reactive ion etching (RIE) approach, buried oxide layer was removed at the same time, and finally, the wafer was backside etched by deep reactive ion etching (DRIE) to release the beam. Moreover, the top and bottom electrode comprises Ti and Pt and was produced with DRIE. On subjecting to base acceleration (2 g), the oscillating cantilever beam produced a load voltage of 1587 mV and output power of 2.099 μW from d_{31} mode across 150 kΩ at a resonant frequency of 255.9 Hz. For d_{33} mode at the resonance of 214 Hz under 2 g base acceleration, an output voltage of 2292 mV and 1.288 μW power was delivered across 510 kΩ load resistance. It has been reported that the poled PZT due to interdigitated electrodes in d_{33} mode results in a nonuniform poling direction, with recorded output power lower as compared to power generation in d_{31} mode.

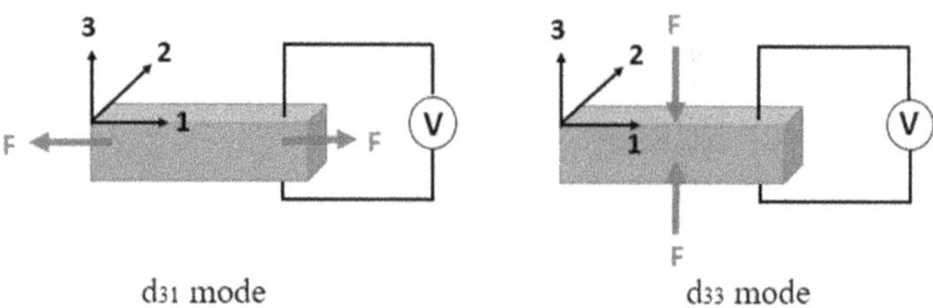

FIGURE 7.3 Illustration of d_{31} and d_{33} operation modes of a piezoelectric material.

A nonlinear cantilever type piezoceramic (PZT-5H), d_{31} mode harvester [77] is developed and experimentally tested. To characterize the PEEH's nonlinear behavior, its bimorph cantilever was fixed to a vibration shaker which consists of two PZT-5H layers charged oppositely. The brass beam was bracketing by PZT-5H layers for providing electrical conductivity between the piezoelectric elements. The harvester was characterized for input excitation frequencies of 530, 535, 540, 545, and 550 Hz under base accelerations of 0.060, 0.145, 0.230, 0.310, 0.430, 0.560, 0.840, 1.12, 1.4, 1.7, and 2 g. A maximum root mean squared (RMS) voltage of 300 mV was obtained at a resonant frequency of 542 Hz across a 100 kΩ resistive load.

Gammaitoni et al. [78] report linear and nonlinear behaviors of PEEH. The harvester has been shown to produce enough power to operate a MEMS device, by harvesting low-frequency (1–100 Hz) nonlinear oscillations; without the need to tune up by altering length, stiffness, or the tip mass attached to the cantilever beam. The harvester, however, requires tuning to adapt to changes in vibration frequency from 80 to 100 Hz while testing for its linear behavior.

W.J. Choi et al. [79] develop a thin-film cantilever type PEEH with a good power density. The harvester, with bimorph (PZT/SiNx) structure having a footprint area of 44,200 μm^2, and a proof mass at the free end, was micro-fabricated. The harvester was modeled for design objectives to predict the effects of the proof mass, damping ratio, and beam shape on power generation. For fabrication, SiNx was first deposited; afterward, the oxide layer was coated on the Silicon (Si) substrate, using plasma-enhanced chemical vapor deposition, followed by ZrO_2 coating using a sol-gel process. A PZT layer was produced at high temperature by precursor gel film pyrolysis process and a layer of Pt/Ti was deposited as the top electrode. SU-8 was then spin-coated, and the structure was released with XeF_2 vapors, and finally patterned into an interdigitated PZT thin film. To harvest low-scale environmental vibrations effectively, a spiral cantilever beam was designed for compactness, lower resonance, and least damping coefficient. It has been shown that the energy harvester is able to generate an output power of 1.01 μW across the terminals of the piezoelectric layer in response to resonance at 150 Hz under 0.5 g. Moreover, across a 5.2 MΩ load resistance, a voltage of 2.4 V has been reported to be produced. The tested device can provide an uninterrupted power supply to an infrastructure monitoring WSN.

Francesco et al. [80] compare the performance of three cantilever-type linear and nonlinear PEEHs of identical dimensions of 1.31 cm^3. The harvesters were installed at three different locations equally apart, at the entrance, center, and exit of a concrete bridge to test the capability of converting the bridge's vibrations into electrical energy. Two of the harvesters were linear, narrowband, while the third energy harvester was nonlinear and bistable broadband. It has been shown experimentally that the harvesters can produce comparatively more power at the entrance of the bridge at 0.36 g base acceleration across a 10 kΩ load. The first linear device resonates at 28 Hz with a bandwidth of 2.5 Hz and can produce an RMS voltage and power of 545.2 mV and 29.72 μW, respectively. The second linear harvester excites at 80.4 Hz with a frequency bandwidth of 4.4 Hz and can generate an RMS voltage of 484.9 mV and power of 23.5 μW. Under the same experimental condition, it has been reported that due to its wide operation frequency bandwidth, the nonlinear harvester was able to produce a voltage of 641 mV and power of 41.0 μW, which are more than that produced by the linear devices of same sizes.

A nonlinear, low-frequency, wideband PEEH is reported in Ref. [81], to demonstrate the feasibility of the model in predicting the device response in low-amplitude vibrations, such as human motion. The harvester produced an output power of 400 μW across a load resistance of 15 kΩ under 1.36 g base accelerations at 7 Hz resonant frequency in the response of a sinusoidal vibration. Under lower base acceleration of 0.7 g, the harvester produced 260 μW power when subjected to periodic vibrations at resonance. Furthermore, at 15 Hz the harvester generates 250 μW power with an efficiency of 15.7% which is less than the power generated in resonance (24.6%).

An AlN-based PEEH was fabricated by micromachining process in Ref. [82]. The harvester comprises a cantilever beam made of Si carrying a proof mass attached to its free end. The device was fabricated with an AlN as a piezoelectric layer and the deposited Al layer and Pt layer were used as the top electrode and bottom electrode. The reported harvester is tested inside the laboratory for a frequency sweep from 201 to 200 Hz under base acceleration from 1 to 8 g. At a resonant frequency of 572 Hz, the harvester generated peak power of 60 μW under 2 g base acceleration.

A PEEH is presented in Ref. [83] for investigating the ultrasonic power transmission for implantable biosensors as it is relatively safe to humans and having less interference with integrated circuits as compared to RF transmission. The harvester can supply power to implantable devices and low-power biosensors through living tissues for monitoring, identification, and treatment of various diseases by harvesting input ultrasonic waves vibration. Metal pads were patterned on the Si layer to stop direct charge penetration and a polyimide protective layer was applied in front of the fabricated model through the DRIE process. The exposed buried oxide was removed using wet hydrofluoric etch to eliminate polyimide protective coat by oxygen plasma at the end. The designed harvester was capable of extracting energy bidirectionally and performed better in the high-frequency range of 38.5–38.7 kHz. When placed at a distance of 0.5 cm from an ultrasonic transducer (biased by 60 V-DC), the harvester produced 21.4 nW power when excited along the x-axis and successfully charged a 1 μF capacitor from 0.51 to 0.95 V in 15 seconds. However, on excitation along Y-axis, the reported PEEH harvested 22.7 nW power under the same experimental conditions.

A cantilever-type nonlinear PEEH has been developed in reference [84]. The device carrying a small mass of 0.8 grams at the free tip of the cantilever has been tested for forward frequency sweep from 5 Hz to 4 kHz. At resonance of 80 Hz and under 0.9 mm beam displacement, the harvester generated a peak load voltage of 1200 mV and a maximum power of 2 μW across a 333 kΩ optimum load.

A thin film, d_{33} mode, PZT-based PEEH has been developed in Ref. [85]. The cantilever-type miniature (44,200 μm^2) device was fabricated, following a three photomasks process which comprises a membrane layer (SiO_2 and/or SiN_x), PZT layer, a ZrO_2 diffusion buffer layer (insulating charge diffusion from piezoelectric layer), a top Pt/Ti electrode, and an SU-8 layer as tip mass. The membrane layer was deposited on Si (100) wafer followed by depositing Pt/Ti as electrode layer and ZrO_2 was deposited on the bottom electrode using a sol-gel spin-on process. PZT layer was deposited on Si, and then PZT and membrane layers were RIE patterned for about an hour with BCl_3:Cl_2, followed by lift-off of Pt/Ti electrode layer, proof mass patterning, and XeF_2 release of the cantilever beam. The cantilever beam exhibits three resonant frequencies at 13.9, 21.9, and 48.5 kHz. At 13.9 kHz, 1 μW power can be generated across 5.2 MΩ, corresponding energy density of 0.74 mWh/cm^2. Moreover, across a 10.1 MΩ resistor with connection to a rectifier circuit, the PEEH can produce 3 V DC voltage output.

Geffrey K. Ottman et al. [86] fabricate a PEEH mainly from AlN and Pt, with an adoptive rectifier circuit attached to the harvester. The harvester was connected to an AC–DC capacitive rectifier; an electrochemical battery and a switch-mode DC-DC converter was also used to control the power supply to the battery. The adaptive control method (DC-DC converter) ensures optimal power transfer and helped in maximizing the stored power in the battery. At an input frequency of 53.8 Hz, the harvester generated an open circuit voltage of 45 V, with a maximum 4.3 mA current at a duty cycle of 3.18%, and the current level maintained consistently above 4 mA at a duty cycle between 2.5% and 4.5%. The harvester can deliver 18 mW power across a resistive load, while when using a 3 V battery, 13 mW power can be stored at optimal duty cycle, giving an estimated power loss of 5 mW. With a resistive load of 24 kΩ connected across the PEEH, load voltage of 20.57 V can be generated, attributing to 18 mW peak power.

Vibration-based Unimorph-type PEEH is presented in Ref. [87], with the device exhibiting three resonant frequencies. The harvester was characterized for forward frequency sweep from 0 to 1400 Hz, and at first resonance, the observed damping ratio was $\zeta_1 = 0.1$. A decrease in voltage was observed as input frequency was increased, and at first resonant frequency of 60.3 Hz, the generated voltage was 7.45 mV. At second (378.6 Hz) and third resonant frequency (1060.6 Hz), the generated voltage levels dropped down to 1.335 and 0.51 mV, respectively.

A low-frequency PEEH is reported by White et al. [88]. The piezo-ceramic efficiency was analyzed in vibration-induced low frequency (below 100 Hz) excitation. Experiments were performed at different input frequencies, and across various load resistances, and it has been found that the output power changes more with changes in input frequency than changing the resistive loads across the harvester. The device was tested for a frequency sweep of 0.1–50 Hz across load resistance from 1 Ω to 100 kΩ, with the highest efficiency found at 5 Hz. Different load resistances of 5.1, 10, 20, 49.9, and 100 kΩ were connected across the harvester at various input frequencies of 1, 2, 6, 10, 14, 20, 30, and 50 Hz to test its performance. A good agreement between the devised model predictions and experiments has been observed for each data set.

Traffic-induced low-frequency and low-amplitude bridge energy harvester is reported by Galchev et al. [89]. Generally, harvesting energy from nonperiodic low-frequency bridge vibrations is challenging due to low excitation levels of 0.01–0.1 g and a low frequency of 1–40 Hz. The cylindrical harvester enclosed tungsten carbide and copper spring. The harvester charged a capacitor from 2 to 2.5 V when excited at 10 Hz while mounted on a vibration shaker. Average power, ranging from 0.46 to 0.72 μW, and peak power of 30–100 μW have been reported from the harvester, with a measured voltage of 18 mV. Two capacitors of 10 μF were used to form a multiplier cascaded stage of 100 μF, in the design.

A cantilever-based PEEH has been developed by Marzencki et al. [90]. The fabrication process involved AlN deposition and patterning over SiO_2 substrate, followed by wet etching. Al was deposited as the top electrode and the electrode was DRIE-etched. Doped Si was used as the bottom electrode, and finally, the bulk Si was etched via DRIE on the backside of the electrode. An overall fabricated harvester of 2 mm^3 was fabricated that delivered 0.8 μW power at a resonant frequency of 300 Hz under 1 g base acceleration.

7.4 COMPARISON AND DISCUSSION

PEEHs can produce higher energy density compared to both EMEHs and ESEHs. The efficiency

$$\eta = U_{\text{out}}/U_{\text{in}} \tag{7.2}$$

of PEEHs reported in the literature may be compared based on output electrical energy in response to the input mechanical excitation [91], where U_{out} is the output electrical energy across load and U_{in} is the input mechanical vibration energy.

Moreover, energy conversion in PEEHs may also be compared by using the analytical model.

$$\lambda_{\max} = k^2/4 - 2k^2 \tag{7.3}$$

presented by Roundy et al. [92]. In model (7.3), k is the coupling coefficient which measures the input excitation conversion efficiency into the output power of an energy harvester and λ is the transmission coefficient and is dependent on load resistance and coupling efficiency. For PEEH according to Roundy et al. [92], the coupling coefficient

$$k^2 = d^2E/\varepsilon \tag{7.4}$$

depends on piezoelectric strain constant d, Young's modulus of elasticity E of material, and ε is the dielectric constant. The maximum power

$$P_{\max} = \lambda_{\max}\omega U_{\text{in}} \tag{7.5}$$

that can be obtained from PEEH is a function of ω, the angular frequency of vibration.

Furthermore, the energy density of the PEEHs can be expressed by Equation (7.6) [92]

$$E_{\max} = k^2\rho\ (QA)^2/4\omega \tag{7.6}$$

where the density of tip-proof mass (on piezoelectric material) is represented by ρ, Q is the quality factor of the harvester, and A is the amplitude of base acceleration.

The electromechanical coupling coefficient of piezoelectric materials depends upon the type of piezoelectric material used and the elastic properties of the suspension system (beam or membrane) used in the energy harvester. The reported PEEHs in literature are compared in Table 7.5 for different parameters such as harvesting mechanism, operation frequency, base acceleration, internal resistance, and peak power values. PEEHs with a proof mass attached to the cantilever beam generate greater power as compared to PEEHs having no tip-mass; this is because

TABLE 7.5
Performance Comparison of the Reported PEEHs

Device Type	Operation Frequency (Hz)	Optimum Load (Ω)	Vibration Intensity (g)	Peak Power (μW)	Refs.
Cantilever-type	255.9	150k	2	2.099	[76]
	150	5.2 M	0.5	1.01	[79]
	28	10k	0.36	29.72	[80]
	7	15k	1.36	400	[81]
Frequency up-conversion type	10	-	0.1	100	[89]
	2	150k	2	43	[93]
	51	-	0.8	0.19	[94]
Membrane type	2.58k	56k	2	1800	[95]
	1.71k	5.6k	-	0.65	[96]
Cantilever type	12	2 M	0.55	30.55	[97]
Frequency up-conversion type	5.6	150k	2	43	[98]
Buckled beam type	3	3.3 M	-	5	[99]
Impact-driven	20k	30k	0.4	51	[100]
Cantilever type	1	1 M	6.3	2779	[101]

the proof mass in a harvester acts as an inertial component due to which relatively high-amplitude oscillation is produced that results in large material deformation (strain) and thus comparatively improved power levels are generated. Similarly, harvesters with larger dimensions produced more power levels at respective resonant frequencies due to bulk piezoelectric material. Base excitation and internal resistance greatly affect the output power of the energy harvesters. PEEHs characterized under vibrations with higher acceleration intensity and having lower internal resistances generated relatively more power as compared to the PEEHs with higher internal impedances. As internal resistance increases, less current can be drawn out of the harvester.

7.5 CONCLUSIONS AND FUTURE RECOMMENDATIONS

Evolutional advances in wireless sensor network (WSNW) technology have been driven by issues arising from battery dependency of WSNs in conjunction with the WSNW. In the last two decades, energy harvesting technology is a much-growing research interest to improve the lifespan of batteries or to reduce its use in WSNs. Piezoelectric materials transform the available mechanical energy, usually, mechanical vibrations in the surroundings to convert it into electrical energy for powering WSNs and other low-power operated devices. These wireless, autonomous, and self-powered devices are fit to be kept in far off and remote locations, such as structural monitoring sensors and GPS tracking devices on objects in the wild. Furthermore, the introduction of nonlinear piezoelectric coupling also adds improvement to the system's performance and better output power generation. Considerable work has also been reported on using different approaches to tune the resonant frequencies of PEEHs using frequency-up-conversion for increasing the frequency range of these harvesters. Harvesting energy from human motion can boost up the performance and reliability of wearable devices by making them highly self-sustained. Moreover, the ultrasonic based wireless transmission energy harvesting is relatively safe to humans, and by this approach, energy can be extracted in two dimensions (X- and Y-axis) which doubles the harvester's efficiency.

REFERENCES

1. Rhimi M, Lajnef N. Tunable energy harvesting from ambient vibrations in civil structures. *J Energy Eng* 2012;138:185–93.
2. Chen J, Qiu Q, Han Y, Lau D. Piezoelectric materials for sustainable building structures: Fundamentals and applications. *Renew Sustain Energy Rev* 2019;101:14–25.
3. Ouyang D, Chen M, Huang Q, Weng J, Wang Z, Wang J. A review on the thermal hazards of the lithium-ion battery and the corresponding countermeasures. *Appl Sci* 2019;9(12): 2483.
4. Pop-Vadean A, Pop PP, Latinovic T, Barz C, Lung C. Harvesting energy an sustainable power source, replace batteries for powering WSN and devices on the IoT. *IOP Conf Ser Mater Sci Eng* 2017;200:0–9.
5. Kyono T, Suzuki RO, Ono K. Conversion of unused heat energy to electricity by means of thermoelectric generation in condenser. *IEEE Trans Energy Convers* 2003;18:330–4.
6. Ackermann T, Soder L. Wind energy technology and current status: A review. *Renew & Sustain Energy Rev* 2000;4:315–74.
7. Narasimhan V, Jiang D, Park SY. Design and optical analyses of an arrayed microfluidic tunable prism panel for enhancing solar energy collection. *Appl Energy* 2016;162:450–9.
8. Kimura S, Sugou T, Tomioka S, Iizumi S, Tsujimoto K, Yasushiro N. Acoustic energy harvester fabricated using sol/gel lead zirconate titanate thin film. *Jpn J Appl Phys* 2011;50:187–90.
9. Reimers CE, Tender LM, Fertig S, Wang W. Harvesting energy from the marine sediment--water interface. *Environ Sci Technol* 2001;35:192–5.
10. Nayyar A, Stoilov V. Power generation from airflow induced vibrations. *Wind Eng* 2015;39:175–82.
11. Khan FU, Izhar. State of the art in acoustic energy harvesting. *J Micromechanics Microengineering* 2015;25:023001.
12. Cook-Chennault KA, Thambi N, Bitetto MA, Hameyie EB. Piezoelectric energy harvesting. *Bull Sci Technol Soc* 2008;28:496–509.
13. Yildiz F. Potential ambient energy-harvesting sources and techniques. *J Technol Stud* 2009;35:40–8.
14. Pareja Aparicio M, Bakkali A, Pelegri-Sebastia J, Sogorb T, Llario V, Bou A. *Radio Frequency Energy Harvesting - Sources and Techniques*. In: Bakkali A, editor. *Renew Energy - Util Syst Integr*, Rijeka: IntechOpen;2016.
15. Jbaily A, Yeung RW. Piezoelectric devices for ocean energy: A brief survey. *J Ocean Eng Mar Energy* 2015;1:101–18.
16. Sue CY, Tsai NC. Human powered MEMS-based energy harvest devices. *Appl Energy* 2012;93:390–403.
17. Purwadi AM, Parasuraman S, Khan MKAA, Elamvazuthi I. Development of Biomechanical Energy harvesting device using heel strike. *Procedia Comput Sci* 2015;76:270–5.
18. Paradiso JA, Starner T. Energy scavenging for mobile and wireless electronics. *IEEE Pervasive Comput* 2005;4:18–27.

19. Zhou M, Al-Furjan MSH, Zou J, Liu W. A review on heat and mechanical energy harvesting from human – Principles, prototypes and perspectives. *Renew Sustain Energy Rev* 2018;82:3582–609.
20. Siddique ARM, Mahmud S, Heyst B Van. A comprehensive review on vibration based micro power generators using electromagnetic and piezoelectric transducer mechanisms. *Energy Convers Manag* 2015;106:728–47.
21. Li H, Tian C, Deng ZD. Energy harvesting from low frequency applications using piezoelectric materials. *Appl Phys Rev* 2014;1:0–20.
22. Paradiso JA, Feldmeier M. A Compact, Wireless, Self-Powered Pushbutton Controller BT - UbiComp 2002: Ubiquitous Computing. UbiComp 2002 Ubiquitous Comput 2001;2201: 299–304.
23. Guo L, Lu Q. Potentials of piezoelectric and thermoelectric technologies for harvesting energy from pavements. *Renew Sustain Energy Rev* 2017;72:761–73.
24. Khan FU, Izhar. Hybrid acoustic energy harvesting using combined electromagnetic and piezoelectric conversion. *Rev Sci Instrum* 2016;87:025003.
25. Wei C, Jing X. A comprehensive review on vibration energy harvesting: Modelling and realization. *Renew Sustain Energy Rev* 2017;74:1–18.
26. Sun C, Shang G, Wang H. On piezoelectric energy harvesting from human motion. *J Power Energy Eng* 2019;07:155–64.
27. Mohamad SH, Thalas MF, Noordin A, Yahya MS, Hassan MHC, Ibrahim Z. A potential study of piezoelectric energy harvesting in car vibration. *ARPN J Eng Appl Sci* 2015;10:8642–7.
28. Roundy S, Wright PK, Rabaey J. A study of low level vibrations as a power source for wireless sensor nodes. *Comput Commun* 2003;26:1131–44. doi:10.1016/S0140-3664(02)00248-7.
29. Iqbal M, Khan FU. Hybrid vibration and wind energy harvesting using combined piezoelectric and electromagnetic conversion for bridge health monitoring applications. *Energy Convers Manag* 2018;172:611–8.
30. Pan Y, Lin T, Qian F, Liu C, Yu J, Zuo J, et al. Modeling and field-test of a compact electromagnetic energy harvester for railroad transportation. *Appl Energy* 2019;247:309–21.
31. Howells CA. Piezoelectric energy harvesting. *Energy Convers Manag* 2009;50:1847–50.
32. Yang Z, Zhou S, Zu J, Inman D. High-performance piezoelectric energy harvesters and their applications. *Joule* 2018;2:642–97.
33. Yang Z, Zu J. High-efficiency compressive-mode energy harvester enhanced by a multi-stage force amplification mechanism. *Energy Convers Manag* 2014;88:829–33.
34. Marzencki M, Basrour S, Charlot B, Grasso A, Colin M, Valbin L. Design and fabrication of piezoelectric micro power generators for autonomous microsystems. *Symp Des Test Integr Packag MEMS/MOEMS* 2005:299–302.
35. Kumar A, Balpande SS, Anjankar SC. Electromagnetic energy harvester for low frequency vibrations using MEMS. *Procedia Comput Sci* 2016;79:785–92.
36. Beeby SP, O'Donnell T. Electromagnetic energy harvesting. 2009. doi:10.1007/978-0-387-76464-1_5.
37. Bakhtiar S, Khan FU. Analytical modeling and simulation of an electromagnetic energy harvester for pulsating fluid flow in pipeline. *Sci World J* 2019;2019, |Article ID 5682517].
38. Khan FU, Iqbal M. Electromagnetic bridge energy harvester utilizing bridge's vibrations and ambient wind for wireless sensor node application. *J Sensors* 2018;2018, |Article ID 3849683].
39. Aljadiri RT, Taha LY, Ivey P. Electrostatic energy harvesting systems: A better understanding of their sustainabilityelectrostatic energy harvesting systems: A better understanding of their sustainability. *J Clean Energy Technol* 2017;5:409–16.

40. Ahmad MR, Md Khir MH, Dennis JO, Zain AM. Fabrication and characterization of the electrets material for electrostatic energy harvester. *J Phys Conf Ser* 2013;476:1–6.
41. Crovetto A, Wang F, Hansen O. Modeling and optimization of an electrostatic energy harvesting device. *J Microelectromechanical Syst* 2014;23:1141–55.
42. Boisseau S, Despesse G, Seddik BA. Electrostatic conversion for vibration energy harvesting. *Small-Scale Energy Harvest* 2012:1–39.
43. Fu Y, Ouyang H, Davis RB. Triboelectric energy harvesting from the vibro-impact of three cantilevered beams. *Mech Syst Signal Process* 2019;121:509–31.
44. Sano C, Mitsuya H, Ono S, Miwa K, Toshiyoshi H, Fujita H. Triboelectric energy harvesting with surface-charge-fixed polymer based on ionic liquid. *Sci Technol Adv Mater* 2018;19:317–23.
45. Berry AL. *The Application of a Triboelectric Energy Harvester in the Packaged Product Vibration Environment*. 2016.
46. Lu W, Xu Y, Zou Y, Zhang L-ao, Zhang J, Wu W, et al. Corrosion-resistant and high-performance crumpled-platinum-based triboelectric nanogenerator for self-powered motion sensing. *Nano Energy* 2020;69:104430.
47. Khan FU, Izhar. Hybrid acoustic energy harvesting using combined electromagnetic and piezoelectric conversion. *Rev Sci Instrum* 2016;87:025003.
48. Khaligh A, Zeng P, Zheng C. Kinetic energy harvesting using piezoelectric and electromagnetic technologiesstate of the art. *IEEE Trans Ind Electron* 2010;57:850–60.
49. Bessaad A, Rhouni A, Basset P, Galayko D. Autonomous CMOS power management integrated circuit for electrostatic kinetic energy harvesters e-KEH. *2019 IEEE Int Conf Electron Circuits Syst*, 2019, pp. 338–41.
50. Khan FU, Iqbal M. Electromagnetic-based bridge energy harvester using traffic-induced bridge's vibrations and ambient wind. *2016 Int Conf Intell Syst Eng ICISE* 2016, 2016, pp. 380–5.
51. Slabov V, Kopyl S, Soares dos Santos MP, Kholkin AL. Natural and eco-friendly materials for triboelectric energy harvesting. *Nano-Micro Lett* 2020;12. doi:10.1007/s40820-020-0373-y.
52. Berdy DF, Valentino DJ, Peroulis D. Kinetic energy harvesting from human walking and running using a magnetic levitation energy harvester. *Sensors Actuators, A Phys* 2015;222:262–71.
53. Wang W, Cao J, Zhang N, Lin J, Liao WH. Magnetic-spring based energy harvesting from human motions: Design, modeling and experiments. *Energy Convers Manag* 2017;132:189–97.
54. Sainthuile T, Delebarre C, Grondel S, Paget C. Vibrational power harvesting for wireless PZT-based SHM applications. *Proc 5th Eur Work - Struct Heal Monit* 2010;2010:679–84.
55. Liang G, Yin Y, Zhang D, Li R, Wu Y, Li Y. Vibration effect produced by raised pavement markers on the exit ramp of an expressway. *J Adv Transp* 2019;2019, lArticle ID 9196303].
56. Matiko JW, Grabham NJ, Beeby SP, Tudor MJ. Review of the application of energy harvesting in buildings. *Meas Sci Technol* 2014;25, 012002.
57. Meninger S, Mur-Miranda JO, Amirtharajah R, Chandrakasan AP, Lang JH. Vibration-to-electric energy conversion. *IEEE Trans Very Large Scale Integr Syst* 2001;9:64–76.
58. Elvin N, Erturk A. Advances in energy harvesting methods. *Adv Energy Harvest Methods*, vol. 9781461457, New York: Springer Science and Business Media; 2013, pp. 1–455.
59. Lee TG, Lee HJ, Kim SW, Kim DH, Han SH, Kang HW, et al. Piezoelectric properties of Pb(Zr, Ti)O_3-Pb(Ni, Nb)O_3 ceramics and their application in energy harvesters. *J Eur Ceram Soc* 2017;37:3935–42.

60. Oh HT, Lee JY, Lee HY. Mn-modified PMN-PZT [Pb(Mg1/3Nb$_2$/3)O$_3$-Pb(Zr, Ti) O$_3$] single crystals for high power piezoelectric transducers. *J Korean Ceram Soc* 2017;54:150–7.
61. Tong S, Ma B, Narayanan M, Liu S, Koritala R, Balachandran U, et al. Lead lanthanum zirconate titanate ceramic thin films for energy storage. *ACS Appl Mater Interfaces* 2013;5:1474–80.
62. Theerthagiri J, Salla S, Senthil RA, Nithyadharseni P, Madankumar A, Arunachalam P, et al. A review on ZnO nanostructured materials: Energy, environmental and biological applications. *Nanotechnology* 2019;30, 392001.
63. Wu WJ, Chen CT, Lin SC, Kuo CL, Wang YJ, Yeh SP. Comparison of the piezoelectric energy harvesters with Si- MEMS and metal-MEMS. *J Phys Conf Ser* 2014;557:0–5.
64. Zvereva OV, Mininzon M, Demianets LN. Hydrothermal growth of OH-free AIPO4 and GaPO4 crystals, the way of twin reducing. *J Phys* 1994;4:19–24.
65. Acosta M, Novak N, Rojas V, Patel S, Vaish R, Koruza J, et al. BaTiO$_3$-based piezoelectrics: Fundamentals, current status, and perspectives. *Appl Phys Rev* 2017;4:041305
66. Ferrari C, Beretta S, Salmaso B, Pareschi G, Tagliaferri G, Basso S, et al. Characterization of ammonium dihydrogen phosphate crystals for soft X-ray optics of the Beam Expander Testing X-ray facility (BEaTriX). *J Appl Crystallogr* 2019;52:599–604.
67. Matsumoto K, Hiruma Y, Nagata H, Takenaka T. Piezoelectric properties of KNbO$_3$ ceramics prepared by ordinary sintering. *Ferroelectrics* 2007;358:169–74.
68. Panda PK, Sahoo B. PZT to lead free piezo ceramics: A review. *Ferroelectrics* 2015;474:128–43.
69. Park S, Kim Y, Jung H, Park JY, Lee N, Seo Y. Energy harvesting efficiency of piezoelectric polymer film with graphene and metal electrodes. *Sci Rep* 2017;7:17290.
70. Fang D, Liu J. Basic equations of piezoelectric materials. *Fract Mech Piezoelectric Ferroelectr Solids*, Berlin, Heidelberg: Springer Berlin Heidelberg; 2013, pp. 77–95.
71. Sodano HA, Inman DJ, Park G. A review of power harvesting from vibration using piezoelectric materials. *Shock Vib Dig* 2004;36:197–205.
72. Von Büren T, Lukowicz P, Tröster G. Kinetic energy powered computing - An experimental feasibility study. *Proc - Seventh IEEE Int Symp Wearable Comput ISWC*, October 21–23, 2003, pp. 22–5.
73. Yang Z, Tan Y, Zu J. A multi-impact frequency up-converted magnetostrictive transducer for harvesting energy from finger tapping. *Int J Mech Sci* 2017;126:235–41.
74. Halvorsen E. Fundamental issues in nonlinear wide-band vibration energy harvesting. *Phys Rev E* 2013;87:042129.
75. Anton SR, Sodano HA. A review of power harvesting using piezoelectric materials (2003–2006). *Smart Mater Struct* 2007;16(3), R1.
76. Wu W, Lee B. *Piezoelectric MEMS Power Generators for Vibration Energy Harvesting.* DOI: 10.5772/51997.
77. Stanton SC, Erturk A, Mann BP, Inman DJ. Nonlinear piezoelectricity in electroelastic energy harvesters: Modeling and experimental identification. *J Appl Phys* 2010;108:1–9.
78. Gammaitoni L, Vocca H, Neri I, Travasso F, Orfei F. Vibration energy harvesting: Linear and nonlinear oscillator approaches (Chapter 7). *Sustain Energy Harvest Technol - Past, Present Futur* 2011:885–90.
79. Choi WJ, Jeon Y, Jeong J-H, Sood R, Kim SG. Energy harvesting MEMS device based on thin film piezoelectric cantilevers. *J Electroceram* 2006;17: 543–548, Springer, USA.
80. Orfei F, Vocca H, Gammaitoni L. Linear and non linear energy harvesting from bridge vibrations. *Paper No: DETC2016-59650, V008T10A054; 8 pages. ASME 2016 International Design Engineering Technical Conferences and Computers and Information in Engineering Conference*, August 21–24, 2016, Charlotte, North Carolina, USA.

81. Andò B, Baglio S, Marletta V, Bulsara AR. Modeling a nonlinear harvester for low energy vibrations. *IEEE Trans Instrum Meas* 2018;68(5):1–9.
82. Elfrink R, Kamel TM, Goedbloed M, Matova S, Hohlfeld D, Van Andel Y, et al. Vibration energy harvesting with aluminum nitride-based piezoelectric devices. *J Micromechanics Microengineering* 2009;19(9), 094005.
83. Zhu Y, Moheimani SOR, Yuce MR. A 2-DOF MEMS ultrasonic energy harvester. *IEEE Sens J* 2011;11:155–61.
84. White NM, Glynne-Jones P, Beeby SP. A novel thick-film piezoelectric micro-generator. *SMART Mater Struct* 2001;10:850–2.
85. Jeon YB, Sood R, Jeong JH, Kim SG. MEMS power generator with transverse mode thin film PZT. *Sensors Actuators, A Phys* 2005;122:16–22.
86. Ottman GK, Hofmann HF, Bhatt AC, Lesieutre GA. Adaptive piezoelectric energy harvesting circuit for wireless remote power supply. *IEEE Trans Power Electron* 2002;17:669–76.
87. Fakhzan MN, Muthalif AGA. Vibration based energy harvesting using piezoelectric material. *2011 4th Int Conf Mechatronics*, 2011, pp. 1–7.
88. Goldfarb M, Jones L. On the efficiency of electric power generation with piezoelectric ceramic. *J Dyn Syst Meas Control - Trans ASME* 1999;121:566.
89. Galchev T, McCullagh J, Peterson RL, Najafi K. Harvesting traffic-induced bridge vibrations. *2011 16th Int Solid-State Sensors, Actuators Microsystems Conf TRANSDUCERS'11*, 2011, pp. 1661–4.
90. Marzencki M, Basrour S, Charlot B. Design, modelling and optimisation of integrated piezoelectric micro power generators. The method design optimization. *Simulation* 2005;3:545–8.
91. Beeby SP, Tudor MJ, White NM. Energy harvesting vibration sources for microsystems applications. *Meas Sci Technol* 2006;17(12), R175.
92. Roundy S. On the effectiveness of vibration-based energy harvesting. *J Intell Mater Syst Struct* 2005;16:809–23.
93. Pillatsch P, Yeatman EM, Holmes AS. A piezoelectric frequency up-converting energy harvester with rotating proof mass for human body applications. *Sensors Actuators, A Phys* 2014;206:178–85.
94. Liu H, Lee C, Kobayashi T, Tay CJ, Quan C. Piezoelectric MEMS-based wideband energy harvesting systems using a frequency-up-conversion cantilever stopper. *Sensors Actuators, A Phys*, vol. 186, Elsevier B.V.; 2012, pp. 242–8.
95. Ericka M, Vasic D, Costa F, Poulin G, Tliba S. Energy harvesting from vibration using a piezoelectric membrane. *J Phys IV JP* 2005;128:187–93.
96. Minazara E, Vasic D, Costa F, Poulin G. Piezoelectric diaphragm for vibration energy harvesting. *Ultrasonics* 2006;44:699–703.
97. Wang W, Cao J, Bowen CR, Zhou S, Lin J. Optimum resistance analysis and experimental verification of nonlinear piezoelectric energy harvesting from human motions. *Energy* 2017;118:221–30.
98. Pillatsch P, Yeatman EM, Holmes AS. A piezoelectric frequency up-converting energy harvester with rotating proof mass for human body applications. *Sensors Actuators, A Phys* 2014;206:178–85.
99. Jiang XY, Zou HX, Zhang WM. Design and analysis of a multi-step piezoelectric energy harvester using buckled beam driven by magnetic excitation. *Energy Convers Manag* 2017;145:129–37.
100. Wei S, Hu H, He S. Modeling and experimental investigation of an impact-driven piezoelectric energy harvester from human motion. *Smart Mater Struct* 2013;22:105020.
101. Izadgoshasb I, Lim YY, Lake N, Tang L, Padilla RV, Kashiwao T. Optimizing orientation of piezoelectric cantilever beam for harvesting energy from human walking. *Energy Convers Manag* 2018;161:66–73.

8 Reverse Engineering the LQR Controller to Find Equivalence with PID Controller

Ali Nasir
University of Central Punjab

CONTENTS

8.1 INTRODUCTION

History of control systems dates to 300 BC involving liquid level control (floats) and flame control in oil lamps. Major progress, however, has occurred after the 17th century with fly ball speed governor, automatic steering of ships, and up till 21st century where we have accomplished the control of spacecraft, airplanes, robots, and many other complex systems.

Two major classes of control are classical control and modern control. The classical control includes transfer function-based (frequency domain) approaches. A transfer function is a ratio of Laplace transform of input and output of a system. A system is anything that is meant for controlling the behavior (or output) of, for example, a car, a pendulum, or an electric circuit. Input is a signal that we supply to the system and output is a signal generated by the system in response to the input that we wish to control. Classical control theory facilitates in determination and analysis of stability, gain margins, phase margins, steady-state error, transient characteristics, and design of proportional-integral-derivative (PID)-type controllers [1,2]. Major tools used in classical control theory are Bode plot, Nyquist plot, Routh Hurwitz stability criteria, and root locus. The goal of controller design in classical control is to achieve some specified stability margins, transient characteristics, and steady-state behavior. The design of control involves deciding how many integral or derivative gains are needed (if any) to achieve the desired specifications and determination of the gain values that ensure the achievement of these specifications.

Modern control on the other hand is built on state-space modeling [3,4]. State-space modeling recognizes that there may be some additional variables in the system other than the input and the output that can play an important role in understanding the systems and designing the controller for the same. These variables are called the internal variables or the state variables of the system. A typical state-space model is a set of first-order linear differential equations (linearity is not required by state space, but this chapter discusses only linear systems). Modern control theory allows us to compute the operating modes of a system, its structural properties (such as controllability and observability), design of controllers by placing closed-loop poles, design of observers, design of dynamic compensators, and optimal control design such as Linear Quadratic Regulator (LQR) [5,6]. LQR is designed based on a specified optimality criterion (or cost function). The cost function involves weighting matrices that determine how expensive the control input is and how expensive it is to have state variables away from their desired value. Appropriate selection of weighting matrices in the LQR design is nontrivial. Unlike PID or any classical controller, LQR does not guarantee any transient specifications such as maximum overshoot, settling time, rise time, time constant, etc.

This chapter develops mathematical relationships between controller gains of LQR and PID for the standard forms of first- and second-order linear time-invariant (LTI) systems. The relationship between the weighting matrices in the cost function of LQR with the closed-loop dynamical properties such as natural frequency, time constant, and damping ratio has been established. Relevant relation between the Q matrix and the location of closed-loop poles has already been investigated [7] and an algorithm has been proposed for finding all possible Q matrices corresponding to any desirable closed-loop pole locations (for second-order LTI systems). Also, the bounds on possible closed-loop pole locations using LQR design have investigated [8]. Similar efforts (the relationship between Q matrix and closed-loop pole locations) have also been made for discrete-time systems [9,10]. Further research regarding the stability robustness achievable by LQR has been carried out [11] and a genetic algorithm-based method has also been proposed for eigenvalue placement of an Linear Quadratic (LQ) optimal system [12]. But there has been a lack of effort in relating the modern control theory with the classical one and determination of the equivalence between the LQR-based and PID-type controllers. Therefore, the original contributions of this chapter are to derive a direct mathematical relationship between the LQR weighting matrices (Q and R) and the corresponding gains of an equivalent PID-type controller. Examples have been provided to demonstrate the trends in the dynamical behavior of closed-loop systems in response to the selection of various LQR cost parameters, hence, bridging the gap between the selection of LQR cost parameters and the resulting overshoot, rise time, settling time, etc.

8.2 MATHEMATICAL PRELIMINARIES

This section describes mathematical models and basic properties associated with first- and second-order LTI systems in classical and modern control theory.

8.2.1 Classical Control

The standard form of a first-order system in classical control is given by

$$G(s) = \frac{1}{\tau s + 1} \tag{8.1}$$

Here, τ is called the time constant. The first-order systems have no overshoot and settling time of a first-order system is roughly 3τ.

The second-order system in classical control has the following standard form:

$$G(s) = \frac{\omega_n^2}{s^2 + 2\zeta\omega_n + \omega_n^2} \tag{8.2}$$

Here, ω_n is the natural frequency of the system, ζ is called the damping ratio, and the product $\zeta\omega_n$ is called the damping factor. Second-order system can be undamped $(\zeta = 0)$, underdamped $(\zeta < 1)$, critically damped $(\zeta = 1)$, over-damped $(\zeta > 1)$, or unstable $(\zeta < 0)$. Time constant for the second-order system is inverse of the damping factor; hence the settling time is roughly $\frac{3}{\zeta\omega_n}$. More precise expression for settling time (for underdamped systems) is as follows:

$$t_s = \begin{cases} \dfrac{3.2}{\zeta\omega_n} & \text{for } 0 < \zeta < 0.7 \\ \dfrac{4.5}{\zeta\omega_n} & \text{for } 0.7 \le \zeta < 1 \end{cases} \tag{8.3}$$

Time to reach the peak for step response of an underdamped second-order system is given by

$$t_{\text{peak}} = \frac{\pi}{\omega_n\sqrt{1-\zeta^2}} \tag{8.4}$$

The corresponding peak value at time $t = t_{\text{peak}}$ is given by

$$y_{\max} = 1 + e^{-\frac{\pi\zeta}{\sqrt{1-\zeta^2}}} \tag{8.5}$$

Consequently, the percentage of overshoot is given by

$$M = 100e^{-\frac{\pi\zeta}{\sqrt{1-\zeta^2}}} \tag{8.6}$$

Notice that the overshoot depends only on ζ. Delay time of a standard second-order underdamped system is the time to reach 50% of the steady-state value and is given by

$$t_d = \frac{\left(1.1 + 0.125\zeta + 0.469\zeta^2\right)}{\omega_n} \tag{8.7}$$

Finally, the rise time (time to reach from 10% to 90% of the steady-state value) of a standard second-order under-damped system is given by

$$t_r = \frac{\left(1 - 0.4167\zeta + 2.917\zeta^2\right)}{\omega_n} \tag{8.8}$$

Now the controller used in classical control is of PID type. The transfer function of a PID controller is given by

$$C(s) = K_P + K_D s + \frac{K_I}{s} = \frac{\left(K_D s^2 + K_P s + K_I\right)}{s} \tag{8.9}$$

Design of controller is essentially equivalent to finding the values of the controller gains (K_P, K_D, K_I) such that the desired specifications (of settling time, rise time, maximum percentage overshoot, etc.) are met. There are many variations of the PID controller, for example, PI, PD, P, PIDD, PDII, and so on.

The closed-loop transfer function with controller and plant is given by

$$G_{CL}(s) = \frac{G(s)C(s)}{1 + G(s)C(s)} \tag{8.10}$$

Feedback loop corresponding to the above equation is shown in Figure 8.2.

8.2.2 Modern Control

This section relates to the discussion of details in the previous section. The state-space system corresponding to the first-order transfer function in (8.1) is given by

$$\begin{aligned} \dot{x} &= -\frac{1}{\tau}x + \frac{1}{\tau}u \\ y &= x \end{aligned} \tag{8.11}$$

Here, x is the only state variable and is equal to the output of the system. The input signal is u. Note that all signals x, y, and u in state space are functions of time.

State-space model for the second-order system corresponding to the transfer function in (8.2) is given as

$$y = \begin{bmatrix} 1 & 0 \end{bmatrix} \begin{bmatrix} x_1 \\ x_2 \end{bmatrix} \tag{8.12}$$

The above system has two state variables, i.e., x_1 and x_2. Here, x_1 is equal to the output of the system and x_2 is equal to the time derivative of x_1.

For controller design in the modern controller theory, a key requirement is controllability. Controllability is the ability to drive the system from any initial state to any final state in finite time using a finite piecewise continuous control signal. Mathematically, there are multiple ways to determine the controllability of a system. Since we are only discussing single-input-single-output (SISO) systems in this chapter, the condition we can use is as follows.

For a SISO system of the form,

$$\begin{aligned} \dot{x} &= Ax + Bu \\ y &= Cx \end{aligned} \tag{8.13}$$

Controllability is guaranteed if and only if the following matrix has nonzero determinant:

$$M = \begin{bmatrix} B & AB & A^2B & \dots & A^{n-1}B \end{bmatrix} \tag{8.14}$$

Here, n is the number of state variables in the model (8.13). Following this definition of controllability, we can conclude that our standard first-order system is controllable if and only if the time constant τ is finite. Similarly, the second-order system is rendered controllable if and only if $\omega_n \neq 0$. From here on, it is assumed that $\tau < \infty$ for first-order systems and $\omega_n \neq 0$ for second-order systems.

The controller under consideration in this chapter is the LQR. LQR is an optimal controller that is used for controlling LTI systems such as the one presented in (8.13). LQR is optimal for the following cost function:

$$J(t_0, x_0, u) = \lim_{T \to \infty} \int_{t_0}^{T} \left(x^T Q x + u^T R u \right) dt \tag{8.15}$$

Here, t_0 is the initial time and x_0 is the initial state of the system to be controlled. Note that the weighting matrices Q and R have the following properties:

$$Q = Q^T \geq 0, R = R^T > 0 \tag{8.16}$$

The optimal feedback control law corresponding to LQR is given by

$$\begin{aligned} u(t) &= -Kx(t) \\ K &= R^{-1} B^T P \end{aligned} \tag{8.17}$$

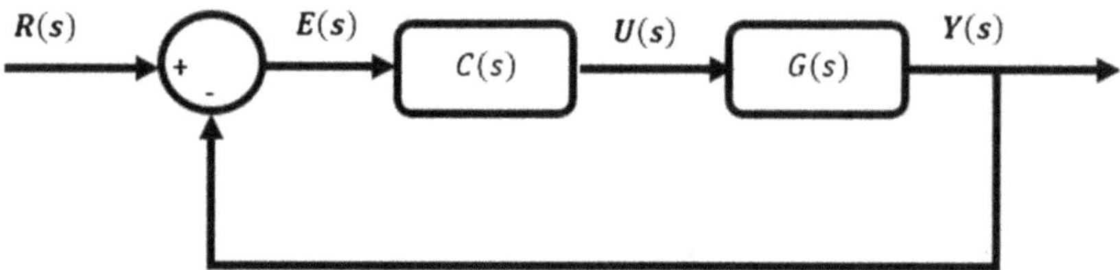

FIGURE 8.1 Feedback loop in classical control.

where P is a positive semidefinite solution to the following algebraic Riccati equation (ARE):

$$PA + A^T P + Q - PBR^{-1}B^T P = 0 \tag{8.18}$$

The above-mentioned LQR is specifically known as infinite horizon LQR. We discuss infinite horizon LQR because it renders the controller gain K as constant, and hence, we can compare it to the gains in the PID controller.

The closed-loop system with LQR controller is given by

$$\begin{aligned} \dot{x} &= (A - BK)x \\ y &= Cx \end{aligned} \tag{8.19}$$

The above-closed loop system is unity feedback (like that in Figure 8.1).

8.3 EQUIVALENCE BETWEEN LQR AND PID

This section presents the mathematical derivations representing the equivalence between LQR and PID controllers under different setups for first- and second-order systems.

8.3.1 First-Order Systems

First-order systems are typically controlled using only a proportional controller (P-controller) or a proportional-integral (PI) controller where removal of steady-state error is required.

8.3.1.1 First-Order System with Proportional Control

The closed-loop transfer function of a first-order system with P-controller is given by

$$G_{CL}(s) = \frac{K_P}{\tau s + K_P + 1} = \frac{\frac{K_P}{\tau}}{s + \frac{(K_P + 1)}{\tau}} \tag{8.20}$$

The corresponding closed-loop characteristic equation of the above system is

$$s + \frac{(K_P + 1)}{\tau} = 0 \tag{8.21}$$

Now, the state-space model of the closed-loop system with LQR is given by

$$\dot{x} = \left(-\frac{1}{\tau} - \frac{1}{\tau}\left(\frac{p}{r\tau} \right) \right) x \tag{8.22}$$

Here, p is the solution of ARE (8.18), r is a scalar version of LQR weighting matrix R, and q is a scalar version of weighting matrix Q and LQR gain is $K = \frac{p}{r\tau}$. The characteristic equation for the closed-loop system in (8.22) is given by

$$s + \frac{1}{\tau} + \frac{1}{\tau}\left(\frac{p}{r\tau} \right) = 0 \tag{8.23}$$

Now, Equation (8.21) is a closed-loop characteristic equation with P-controller, and Equation (8.23) is a closed-loop characteristic equation with LQR. For equivalence, we compare coefficients and obtain

$$\left(1 + \frac{p}{r\tau} \right) = 1 + K_P \tag{8.24}$$

Above equation implies that

$$K_P = \frac{p}{r\tau} \tag{8.25}$$

Equation (8.25) reveals that P-controller gain is to be exactly equal to that of LQR for equivalence. This means that if a P-controller is designed using (8.25) instead of transient and steady-state specifications, it is essentially equivalent to an LQR controller. Vice versa, if an LQR has been designed, one can find the corresponding transient and steady-state characteristics with root locus, bode plot, etc., by equating the gain.

The relation (8.25) is further investigated by solving the underlying Riccati equation for the positive root of p which is given by

$$p = r\tau \left(-1 + \sqrt{1 + \frac{q}{r}} \right) \tag{8.26}$$

Substituting (8.26) in (8.25), we get

$$K_P = -1 + \sqrt{1 + \frac{q}{r}} \tag{8.27}$$

Using (8.27), one can relate the cost parameters of LQR, i.e., q and r directly with K_P.

Example 1

Consider the following first-order system:

$$G(s) = \frac{1}{s+1}$$

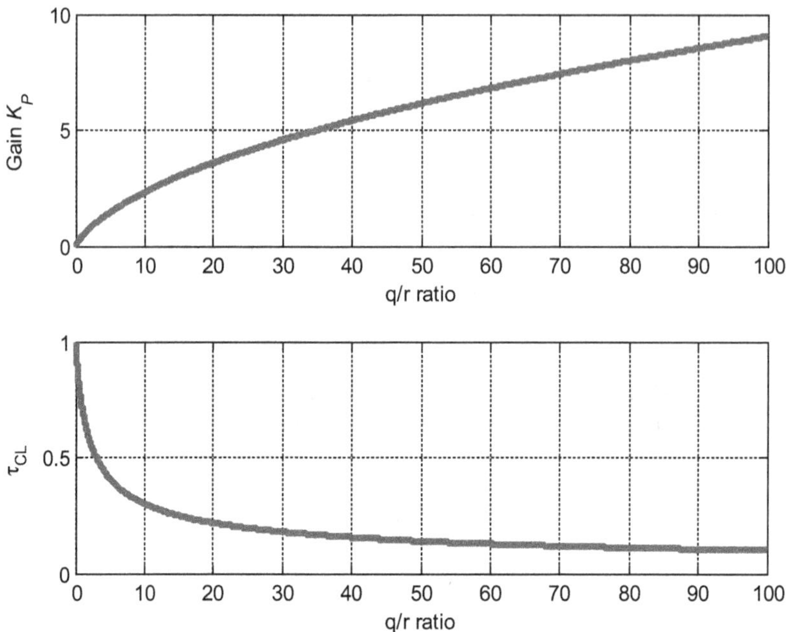

FIGURE 8.2 First-order system with proportional control.

If a P-controller is applied to this system, using the LQR approach, how would the selection of q and r affect the time constant of the resulting closed-loop system? How would the gain be affected by the selection of parameters?

Solution

The closed-loop system with proportional control is given by

$$G_{CL}(s) = \frac{K_P}{\dfrac{s}{1+K_P}+1}$$

Hence, the time constant of the closed-loop system is

$$\tau_{CL} = \frac{1}{1+K_P}$$

Next, we use the above expression along with (8.27) to get the trends of τ_{CL} and K_P with respect to the ratio q/r (see Figure 8.2). It is evident that the gain increases while τ_{CL} decreases as the q/r ratio is increased.

8.3.1.2 First-Order System with PI Control

In the case of the PI controller, the closed-loop transfer function of the feedback system becomes

$$G_{CL}(s)=\frac{(K_P s+K_I)}{\tau s^2+s+K_P s+K_I}=\frac{\frac{(K_P s+K_I)}{\tau}}{s^2+s\frac{(1+K_P)}{\tau}+\frac{K_I}{\tau}} \tag{8.28}$$

The corresponding characteristic equation is given by

$$s^2+s\frac{(1+K_P)}{\tau}+\frac{K_I}{\tau}=0 \tag{8.29}$$

On the other hand, implementation of integral control with LQR requires the addition of a new state called integrator state. So, with integrator state, the system to be controlled is given by

$$\begin{bmatrix} \dot{x}_1 \\ \dot{x}_2 \end{bmatrix}=\begin{bmatrix} -\frac{1}{\tau} & 0 \\ 1 & 0 \end{bmatrix}\begin{bmatrix} x_1 \\ x_2 \end{bmatrix}+\begin{bmatrix} \frac{1}{\tau} \\ 0 \end{bmatrix}u$$
$$y=\begin{bmatrix} 1 & 0 \end{bmatrix}\begin{bmatrix} x_1 \\ x_2 \end{bmatrix} \tag{8.30}$$

Here, x_2 is the integrator state since its derivative is equal to the output y. Now, the closed-loop system with LQR applied to the system in (8.30) is as follows:

$$\dot{x}=\begin{bmatrix} -\frac{1}{\tau}-\frac{p_1}{r\tau^2} & -\frac{p_2}{r\tau^2} \\ 1 & 0 \end{bmatrix}x \tag{8.31}$$

where the underlying matrices related to LQR are

$$R=r, Q=\begin{bmatrix} q_1 & q_2 \\ q_2 & q_3 \end{bmatrix}, P=\begin{bmatrix} p_1 & p_2 \\ p_2 & p_3 \end{bmatrix}, K=\begin{bmatrix} \frac{p_1}{r\tau} & \frac{p_2}{r\tau} \end{bmatrix} \tag{8.32}$$

Now, the closed-loop characteristic equation is given by

$$\det\left(sI-(A-BK)\right)=s^2+s\left(\frac{1}{\tau}+\frac{p_1}{r\tau^2}\right)+\frac{p_2}{r\tau^2}=0 \tag{8.33}$$

Comparing the coefficients of (8.33) and (8.29), the result is:

$$\frac{(1+K_P)}{\tau}=\left(\frac{1}{\tau}+\frac{p_1}{r\tau^2}\right)\rightarrow K_P=\frac{p_1}{r\tau} \tag{8.34}$$

and

$$\frac{K_I}{\tau}=\frac{p_2}{r\tau^2}\rightarrow K_I=\frac{p_2}{r\tau} \tag{8.35}$$

Note that the equations (8.34) and (8.35) provide a way to correlate the cost of control input in modern control theory with transient and steady-state specifications in classical control theory. Solving the Riccati equation for p_1 and p_2 in this case yields

$$p_1 = \tau\left(-r + 2\sqrt{r^2 + r\left(2\tau\sqrt{rq_3} + q_1\right)}\right) \tag{8.36}$$

and

$$p_2 = \tau\sqrt{rq_3} \tag{8.37}$$

It is interesting to note that the element q_2 in the cost matrix, Q, of LQR has an impact on the gains K_P and K_I. Similarly, the element p_3 of the solution matrix of Riccati equation does not appear in the equations of K_P and K_I. Another interesting thing here to notice is the relationship between the controller gain, i.e.,

$$K_P = \frac{1}{r}\left(-r + 2\sqrt{r^2 + r\left(2\tau r K_I + q_1\right)}\right) \tag{8.38}$$

The above result is interesting because it converts the problem of finding two separate gains into a problem of finding only one gain (the other gain follows from the relationship). This situation is like a common practice in classical control where the designer tries to fix the ratio of K_P and K_I to place the zero of PI controller at an appropriate location.

Example 2

Consider the following first-order system:

$$G(s) = \frac{1}{0.1s + 1}$$

If a PI control is applied to this system, using the LQR approach, how would the selection of q and r affect the dynamics of the resulting closed-loop system? How would the gains be affected by the selection of parameters?

Solution

The closed-loop system with proportional control is given by

$$G_{CL}(s) = 10\left(\frac{K_P s + K_I}{s^2 + \dfrac{s(1 + K_P)}{0.1} + \dfrac{K_I}{0.1}}\right)$$

Since the closed-loop system is a second-order system, we approximate its natural frequency $(\omega_n)_{CL}$ and damping ratio (ζ_{CL}) as

$$(\omega_n)_{CL} = \sqrt{10K_I}\ ,\ \zeta_{CL} = \frac{(1 + K_P)}{2\sqrt{0.1K_I}}$$

In this example, the Q matrix of LQR has two important parameters, i.e., q_1 and q_3. Therefore, we calculate the effects of q_1, q_3, and r separately as shown below (see Figures 8.3–8.5). Figure 8.3 shows the effect of parameter q_1 on the gain K_P and closed-loop damping ratio. It is evident that the gain and the damping ratio are increased by increasing q_1. This implies that q_1 helps in making the system more and more overdamped. Note that q_1 has no effect on closed-loop natural frequency or the gain K_I. For the results in Figure 3.2, the values of r and q_1 have been fixed at the value 10. Next, Figure 8.4 shows the effects of parameter q_3 on the controller gains, natural frequency, and damping ratio. It is evident that both gains and the natural frequency of the closed-loop system increase with q_3 whereas the damping ratio decreases. This result is interesting because it indicates that two different elements of the matrix Q may have opposite effects on closed-loop dynamics. An explanation of why q_3 reduces the damping of the closed-loop system is that it increases the integral gain which is known to worsen the transient behavior of the closed-loop system. For the results in Figure 8.4, the values of q_1 and r have been set at 10. Finally, Figure 8.5 shows the trends in the gains and closed-loop dynamics with respect to parameter r. It is evident that the natural frequency, damping ratio, and the gains decrease with the increase in r. Values of q_1 and q_3 have been set at 10 for the results in Figure 8.5.

Note that the settling time, rise time, overshoot, and other quantities can be calculated from ζ_{CL} and $(\omega_n)_{CL}$.

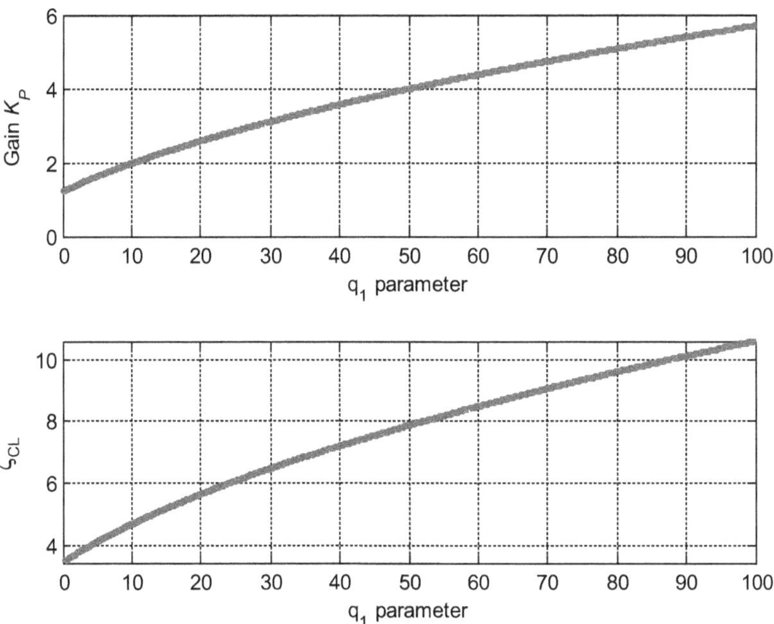

FIGURE 8.3 First-order system with PI control, effects of q_1 on gain, and closed-loop dynamics.

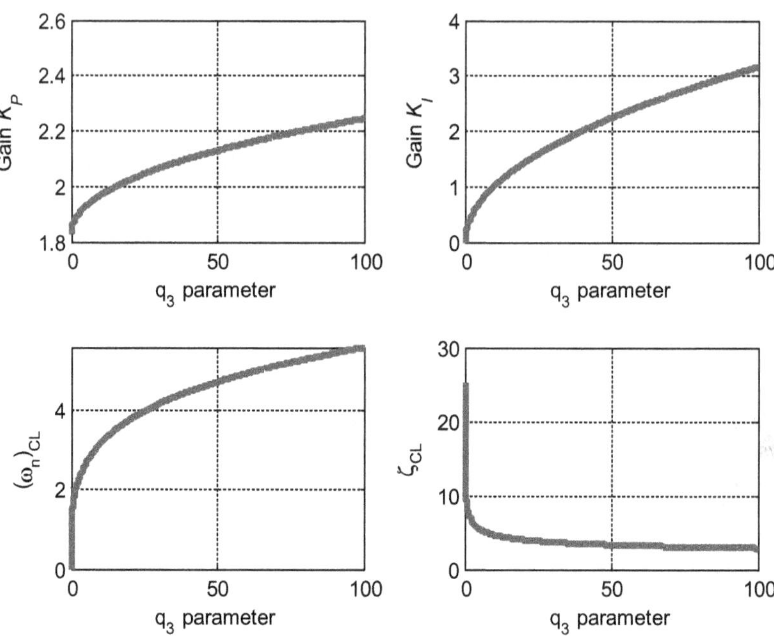

FIGURE 8.4 First-order system with PI control, effects of q_3 on gain, and closed-loop dynamics.

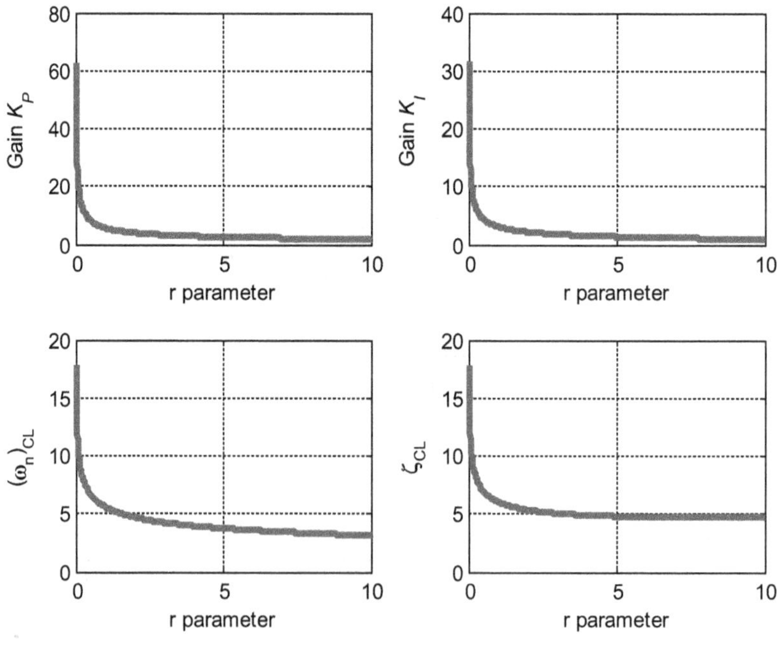

FIGURE 8.5 First-order system with PI control, effects of r on gain, and closed-loop dynamics.

8.3.2 Second-Order Systems

Classical second-order systems show a variety of behavior such as overdamped, underdamped, undamped, and critically damped. Therefore, we present more controller types for second-order systems in this section compared to the first-order systems in the previous section.

8.3.2.1 Second-Order Systems with Proportional Control

The closed-loop transfer function of a standard second-order system (8.2) with a P-controller is given by

$$G_{CL}(s) = \frac{K_P\omega_n^2}{s^2 + 2\zeta\omega_n s + K_P\omega_n^2 + \omega_n^2} \tag{8.39}$$

The corresponding characteristic equation is given by

$$s^2 + 2\zeta\omega_n s + K_P\omega_n^2 + \omega_n^2 = 0 \tag{8.40}$$

Corresponding closed-loop state-space system is given by:

$$\begin{aligned} \dot{x} &= \begin{bmatrix} 0 & 1 \\ -\omega_n^2 & -2\zeta\omega_n \end{bmatrix} \begin{bmatrix} x_1 \\ x_2 \end{bmatrix} + \begin{bmatrix} 0 \\ \omega_n^2 \end{bmatrix} \left[\begin{bmatrix} k_1 & k_2 \end{bmatrix} \begin{bmatrix} x_1 \\ x_2 \end{bmatrix} \right] \\ y &= \begin{bmatrix} 1 & 0 \end{bmatrix} \begin{bmatrix} x_1 \\ x_2 \end{bmatrix} \end{aligned} \tag{8.41}$$

The above equation can be rewritten in terms of LQR as follows:

$$\dot{x} = \begin{bmatrix} 0 & 1 \\ -\omega_n^2 - \omega_n^2 k_1 & -2\zeta\omega_n - \omega_n^2 k_2 \end{bmatrix} x \tag{8.42}$$

The underlying LQR-related matrices and parameters are

$$R = r, Q = \begin{bmatrix} q_1 & q_2 \\ q_2 & q_3 \end{bmatrix}, P = \begin{bmatrix} p_1 & p_2 \\ p_2 & p_3 \end{bmatrix}, \begin{bmatrix} k_1 & k_2 \end{bmatrix} = \begin{bmatrix} \frac{\omega_n^2 p_2}{r} & \frac{\omega_n^2 p_3}{r} \end{bmatrix} \tag{8.43}$$

Now, the closed-loop characteristic equation with LQR control is given by

$$\det(sI - (A - BK)) = s^2 + s\left(2\zeta\omega_n + \omega_n^2 k_2\right) + \omega_n^2 + \omega_n^2 k_1 = 0 \tag{8.44}$$

Comparing coefficients of (8.40) and (8.44), the result is

$$\left(2\zeta\omega_n + \omega_n^2 k_2\right) = 2\zeta\omega_n \rightarrow k_2 = 0 \tag{8.45}$$

and

$$\omega_n^2 + \omega_n^2 k_1 = K_P \omega_n^2 + \omega_n^2 \rightarrow K_P = k_1 = \frac{\omega_n^2 p_2}{r} \tag{8.46}$$

Equations (8.45) and (8.46) indicate that for equivalence between LQR and a P-controller, in this case, one of the gain values of LQR control must be zero and the second values must equal that of K_P. Furthermore, the expression of p_2 is as follows:

$$p_2 = \frac{1}{\omega_n^2}\left(-r + \sqrt{r^2 + q_1 r}\right) \tag{8.47}$$

Substituting the above expression into (8.46) gives the following relationship between K_P and LQR weighting parameter:

$$K_P = \left(-1 + \sqrt{1 + \frac{q_1}{r}}\right) \tag{8.48}$$

Notice the resemblance between (8.27) and (8.48).

Example 3

Consider the following second-order system:

$$G(s) = \frac{0.01}{s^2 + 0.1s + 0.01}$$

If a proportional control is applied to this system, using the LQR approach, how would the selection of q and r affect the dynamics of the resulting closed-loop system? How would the gains be affected by the selection of parameters?

Solution

The closed-loop system with proportional control is given by

$$G(s) = \frac{0.01K_P}{s^2 + 0.1s + 0.01 + K_P}$$

Now, the natural frequency and damping ratio of the closed-loop system are as follows:

$$(\omega_n)_{CL} = \sqrt{0.01 + K_P}$$

$$\zeta_{CL} = \frac{0.05}{\sqrt{0.01 + K_P}}$$

Next, we use the relation (8.48) and the above relations to determine the following trends (see Figure 8.6). It is evident that the damping ratio decreases whereas the natural frequency and the gain increase with the ration q_1/r. It is interesting to note that the damping factor $\left((\omega_n)_{CL}\zeta_{CL}\right)$ does not change significantly with respect to q_1/r in this case.

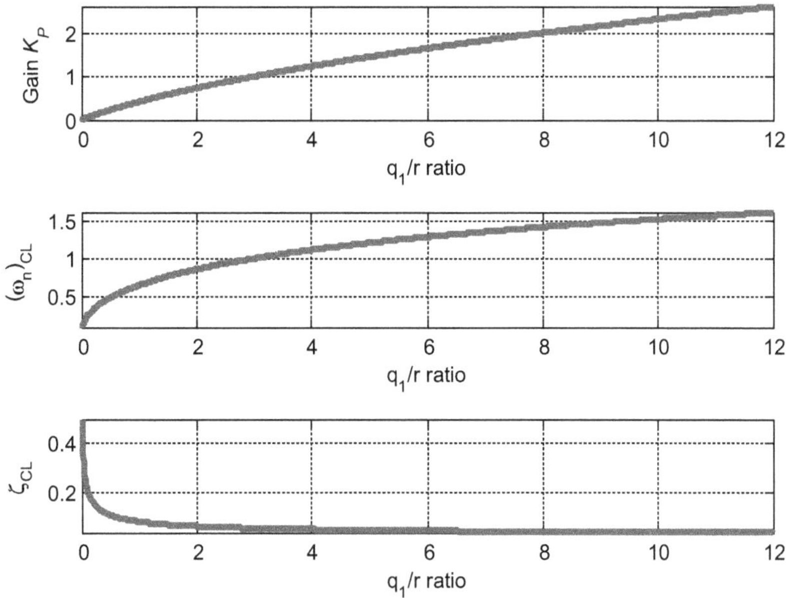

FIGURE 8.6 Second-order system with proportional control.

8.3.2.2 Second-Order System with Proportional-Derivative Control

The closed-loop transfer function of a standard second-order system with proportional-derivative control is given by

$$G_{CL}(s) = G_{CL}(s) = \frac{K_D\omega_n^2 s + K_P\omega_n^2}{s^2 + \left(2\zeta\omega_n + K_D\omega_n^2\right)s + \left(K_P\omega_n^2 + \omega_n^2\right)} \tag{8.49}$$

Hence, the closed-loop characteristic equation is given by

$$s^2 + \left(2\zeta\omega_n + K_D\omega_n^2\right)s + \left(K_P\omega_n^2 + \omega_n^2\right) = 0 \tag{8.50}$$

The corresponding state-space system, in this case, is the same as in (8.42) above and the corresponding closed-loop characteristic equation is the same as in (8.44). Therefore, comparing coefficients of (8.44) and (8.50) leads to

$$2\zeta\omega_n + K_D\omega_n^2 = 2\zeta\omega_n + \omega_n^2 k_2 \rightarrow K_D = k_2 = \frac{\omega_n^2 p_3}{r} \tag{8.51}$$

and

$$K_P\omega_n^2 + \omega_n^2 = \omega_n^2 + \omega_n^2 k_1 \rightarrow K_P = k_1 = \frac{\omega_n^2 p_2}{r} \tag{8.52}$$

Further calculations of p_2 and p_3 lead to the K_P as in (8.48) and K_D as given below:

$$K_D = \frac{1}{\omega_n}\left(2\zeta + \sqrt{4\zeta^2 - \frac{\omega_n^2}{r}\left(q_3 + \frac{2r}{\omega_n^2}\left(-1 + \sqrt{1 + \frac{q_1}{r}}\right)\right)}\right) \tag{8.53}$$

Note that q_2 of the Q matrix in -LQR has no role in defining K_P or K_D. Also, note that K_D and K_P are related to each other because (8.53) and (8.48) lead to the following expression for K_D:

$$K_D = \frac{1}{\omega_n}\left(2\zeta + \sqrt{4\zeta^2 - \frac{\omega_n^2}{r}\left(q_3 + \frac{2r}{\omega_n^2} K_P\right)}\right) \tag{8.54}$$

The above relation is like a common practice in classical control where the designer of the controller sometimes fixes the ratio between K_P and K_D to place the zero of the PD controllers at an appropriate location.

Example 4

Consider the following second-order system:

$$G(s) = \frac{0.01}{s^2 + 0.1s + 0.01}$$

If a proportional-derivative control is applied to this system, using the LQR approach, how would the selection of q and r affect the dynamics of the resulting closed-loop system? How would the gains be affected by the selection of parameters?

Solution

The closed-loop system with proportional control is given by

$$G(s) = \frac{0.01(K_P + K_D s)}{s^2 + (0.1 + 0.01K_D)s + 0.01 + 0.01K_P}$$

Now, the natural frequency and damping ratio of the closed-loop system are as follows:

$$(\omega_n)_{CL} = 0.1\sqrt{1 + K_P}$$

$$\zeta_{CL} = \frac{(1 + 0.1K_D)}{2\sqrt{1 + K_P}}$$

Using (8.48), (8.54), and above relations of $(\omega_n)_{CL}$ and ζ_{CL}, we obtain the following trends (see Figures 8.7–8.9). Graphs in Figure 8.7 indicate that the proportional gain increases with q_1 whereas the derivative gain decreases with q_1. Furthermore, the natural frequency of the closed-loop system increases whereas the closed-loop damping ratio decreases with q_1. The values of r and q_3 have been set at 10 in these results. Figure 8.8 shows that both the derivative gain and the closed-loop

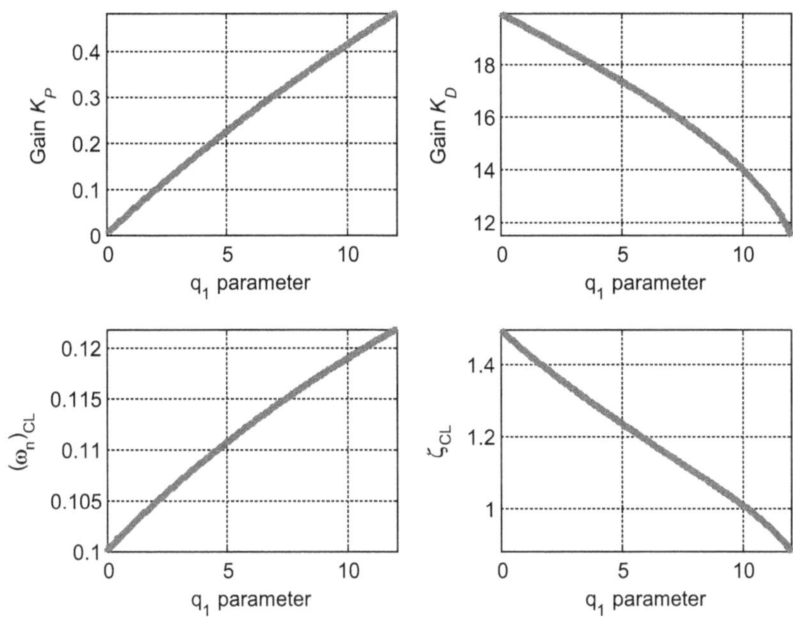

FIGURE 8.7 Second-order system with PD control, effects of q_1.

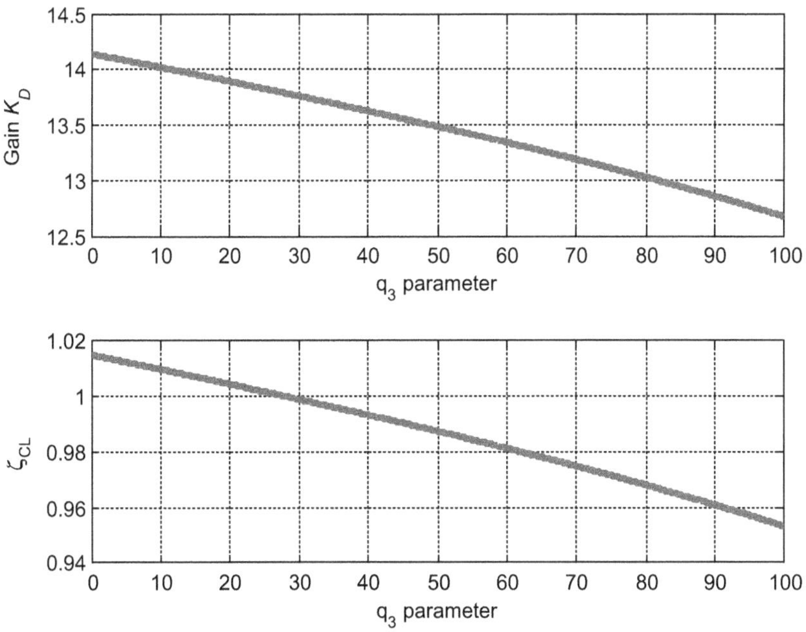

FIGURE 8.8 Second-order system with PD control, effects of q_3.

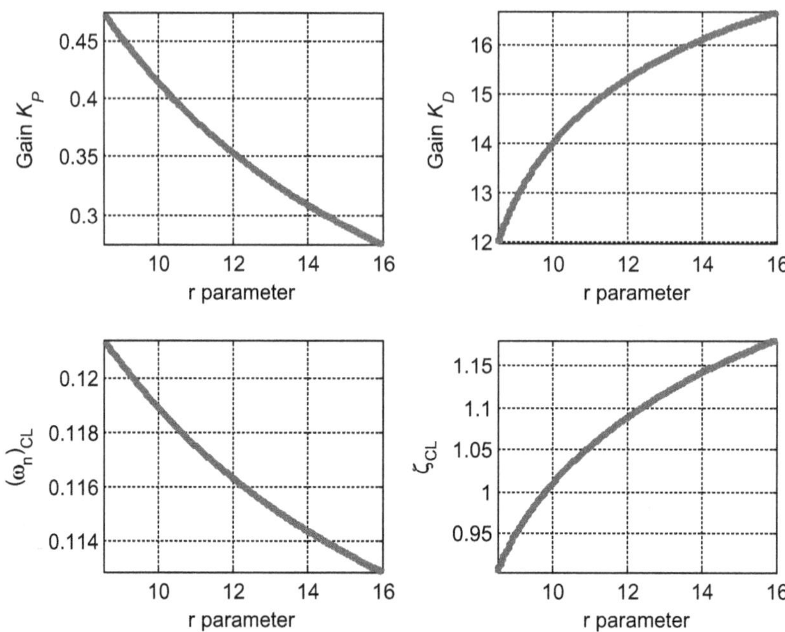

FIGURE 8.9 Second-order system with PD control, effects of *r*.

damping ratio decrease with q_3. Hence q_1 and q_3 have similar effect on the derivative gain and the closed-loop damping ratio in this case. Note that q_3 does not affect the proportional gain or the closed-loop natural frequency. Values of q_1 and r in these results have been set at 10. Finally, Figure 8.9 shows that r has opposite effects of the gains and closed-loop dynamics compared with that of q_1.

8.3.2.3 Second-Order System with PID Control

The closed-loop transfer function of a standard second-order system with PID control is given by

$$\frac{\omega_n^2\left(K_D s^2 + K_P s + K_I\right)}{s^3 + s^2\left(2\zeta\omega_n + \omega_n^2 K_D\right) + s\left(\omega_n^2 K_P + \omega_n^2\right) + \omega_n^2 K_I} \tag{8.55}$$

The corresponding closed-loop characteristic equation is given by

$$s^3 + s^2\left(2\zeta\omega_n + \omega_n^2 K_D\right) + s\left(\omega_n^2 K_P + \omega_n^2\right) + \omega_n^2 K_I = 0 \tag{8.56}$$

Now, the state space system requires an additional state due to the involvement of integral control. Hence the corresponding state-space model is given by

$$\dot{x} = \begin{bmatrix} 0 & 1 & 0 \\ -\omega_n^2 & -2\zeta\omega_n & 0 \\ 1 & 0 & 0 \end{bmatrix} \begin{bmatrix} x_1 \\ x_2 \\ x_3 \end{bmatrix} + \begin{bmatrix} 0 \\ \omega_n^2 \\ 0 \end{bmatrix} u$$

$$y = \begin{bmatrix} 1 & 0 & 0 \end{bmatrix} \begin{bmatrix} x_1 \\ x_2 \\ x_3 \end{bmatrix} \tag{8.57}$$

Condition for controllability of the above model is $\omega_n \neq 0$ which is already our standing assumption. The corresponding closed-loop system model with LQR is given by

$$\dot{x} = \begin{bmatrix} 0 & 1 & 0 \\ -\omega_n^2 - \frac{p_2\omega_n^4}{r} & -2\zeta\omega_n - \frac{p_4\omega_n^4}{r} & -\frac{p_5\omega_n^4}{r} \\ 1 & 0 & 0 \end{bmatrix} x \tag{8.58}$$

The LQR-related matrices and parameters underlying the above model are as follows:

$$Q = \begin{bmatrix} q_1 & q_2 & q_3 \\ q_2 & q_4 & q_5 \\ q_3 & q_5 & q_6 \end{bmatrix}, R = r, P = \begin{bmatrix} p_1 & p_2 & p_3 \\ p_2 & p_4 & p_5 \\ p_3 & p_5 & p_6 \end{bmatrix}, K = \begin{bmatrix} \frac{p_2\omega_n^2}{r} & \frac{p_4\omega_n^2}{r} & \frac{p_5\omega_n^2}{r} \end{bmatrix} \tag{8.59}$$

Next, the closed-loop characteristic equation is given by

$$\det\left(sI - (A - BK)\right) = s^3 + s^2\left(2\zeta\omega_n + \frac{p_4\omega_n^4}{r}\right) + s\left(\omega_n^2 + \frac{p_2\omega_n^4}{r}\right) + \frac{p_5\omega_n^4}{r} = 0 \tag{8.60}$$

Comparing coefficients of (8.60) and (8.56), the result is

$$2\zeta\omega_n + \omega_n^2 K_D = 2\zeta\omega_n + \frac{p_4\omega_n^4}{r} \rightarrow K_D = \frac{p_4\omega_n^2}{r} \tag{8.61}$$

Similarly,

$$K_P = \frac{p_2\omega_n^2}{r} \tag{8.62}$$

and

$$K_I = \frac{p_5\omega_n^2}{r} \tag{8.63}$$

After solving the Riccati equation, we obtain the following expressions for the gains:

$$K_I = \sqrt{\frac{q_6}{r}} \tag{8.64}$$

$$K_P = \frac{\omega_n^2}{r}\left(-\frac{b}{4a} + S + \frac{1}{2}\sqrt{-4S^2 - 2f - \frac{g}{S}}\right),$$

$$f = \frac{\left(8ac - 3b^2\right)}{8a^2},$$

$$g = \frac{\left(b^3 - 4abc + 8a^2 d\right)}{8a^3},$$

$$S = \frac{1}{2}\sqrt{-\frac{2}{3}f + \frac{1}{3a}\left(h + \frac{\Delta_0}{h}\right)},$$

$$h = \sqrt[3]{\frac{\left(\Delta_1 + \sqrt{\Delta_1^2 - 4\Delta_0^3}\right)}{2}},$$

$$\Delta_0 = c^2 - 3bd + 12ae,$$

$$\Delta_1 = 2c^3 - 9bcd + 27b^2 e + 27ad^2 - 72ace,$$

$$a = \frac{\omega_n^8}{r^2},$$

$$b = \frac{4\omega_n^6}{r},$$

$$c = 4\omega_n^4\left(1 + \frac{q_5}{r}\right) - 2\omega_n^3\left(8\zeta\sqrt{\frac{q_6}{r}} + \frac{2\omega_n q_1}{r}\right),$$

$$d = 4\omega_n^2\left(2q_5 - q_1\right) - 8\left(q_6 + 4\omega_n r\zeta\sqrt{\frac{q_6}{r}}\right),$$

$$e = 4q_5^2 + q_1^2 - 4q_1q_5 - 4q_6q_4 + 60\left(\frac{q_6\zeta^2 r}{\omega_n^2}\right) + \frac{16r\zeta}{\omega_n}\sqrt{\frac{q_6}{r}}\left(q_1 - 2q_5\right). \quad (8.65)$$

Notice the appearance of the right-hand side of (8.64) in different coefficients involved in the expression for K_P in (8.65). Similarly, K_D can be written in terms of K_P as follows:

$$K_D = \frac{1}{\omega_n r}\left(2\zeta r + \sqrt{\zeta^2 r^2 + q_4 r\omega_n^2 + 2r^2 K_P}\right) \quad (8.66)$$

Hence all three gains are related to each other.

8.4 DISCUSSION AND CONCLUSIONS

This chapter has presented a way to connect the classical control design specifications (such as rise time, settling time, percentage overshoot, etc.) with the cost parameters in LQR-based modern control. Mathematical relations between the classical control gains and the cost parameters in LQR have been derived. The way to use the derived relations for determination of the closed-loop dynamical properties resulting from chosen cost parameters has been shown through examples.

The scope of this chapter is the first- and second-order standard LTI systems. Generalization of the presented concepts for time-varying systems, nonlinear systems, or even higher order LTI systems is highly nontrivial. A hint of the challenges involved in higher-order generalization is available in Section 8.3.2.3 where the relationships are quite complex and some parameter values may lead to the imaginary values of the gains.

Despite the limited scope, the results of this chapter can be extended for nonstandard first and second-order systems with the help of precompensators. Also, the results can be used for approximating the behavior of higher order systems that involve only one or two dominant poles and all other poles have only short-term transients, i.e., those poles are farther away on the left half of the complex plane. Another application of the study conducted in this chapter is in the determination of the relationship between the closed-loop bandwidth of the system with the LQR gain parameters. Standard equations relating bandwidth to the natural frequency and damping ratio can be found in any of the standard control systems textbooks, e.g., those mentioned in the list of references.

The concept has applications among others, especially in the control of strategic and nonstrategic machine tools.

REFERENCES

1. Kuo BC. *Automatic Control Systems*. Prentice Hall PTR; 1987 Jan 1, England, London.
2. Nise Norman S. *Control Systems Engineering*, (With CD). John Wiley & Sons; 2007, Hoboken, New Jersey, United States.
3. Bishop RH. *Modern Control Systems Analysis and Design Using MATLAB and SIMULINK*. Addison-Wesley Longman Publishing Co., Inc.; 1996 Dec 1, Boston, Massachusetts, United States.
4. Chen CT. *Linear System Theory and Design*. Oxford University Press, New York; 1998.
5. Vargas AN, Ishihara JY, do Val JB. Linear quadratic regulator for a class of Markovian jump systems with control in jumps. In *49th IEEE Conference on Decision and Control (CDC)*; 2010 Dec 15 (pp. 2282–2285). IEEE.
6. Dorato P, Cerone V, Abdallah C. *Linear-Quadratic Control: An Introduction*. Simon & Schuster, Inc.; 1994 Aug 1.
7. Di Ruscio D, Balchen Jens G. A Schur method for designing LQ-optimal systems with prescribed eigenvalues. *Modeling, Identification and Control Journal,* 11(1), 1990:55–72.
8. Di Ruscio D. On the location of LQ-optimal closed-loop poles. *Proceedings of the 30th IEEE Conference on Decision and Control*; 1991 (pp. 2317–2320, vol. 3), IEEE (Institute of Electrical and Electronics Engineers), Brighton, UK.

9. Mansouri N. The boundary of the discrete LQ optimal closed-loop poles. In *2003 4th International Conference on Control and Automation Proceedings*; 2003 Jun 12 (pp. 355–359). IEEE.
10. Khaloozadeh H. A new result for closed-loop poles region of LQ-optimal discrete-time systems. *Optimal Control Applications and Methods*; 2008 Mar;29(2):85–99.
11. Di Ruscio D. Measures for stability robustness in linear quadratic systems. In *[1992] Proceedings of the 31st IEEE Conference on Decision and Control*; 1992 Dec 16 (pp. 1554–1556). IEEE, New York City, New York, United States.
12. Mansouri N, Khaloozadeh H. The GA approach to the eigenvalue placement of an LQ optimal system. In *Proceedings 2002 IEEE International Conference on Artificial Intelligence Systems (ICAIS 2002)*; 2002 Sep 5 (pp. 340–344). IEEE.

Section 3

Machine Tools and Computational Analysis

This section covers work related to sustainable development using finite analysis, machine vision in sheet metal, and power consumption during machine tool operatiom. These topics are included based on their comparative merit and provide an in-depth knowledge of processes involved in system engineering design, operation, and monitoring of machine tools. A chapter, aparently away from the main theme of the presented work, is included to emphasize the effect of emmission and pollution caused by the machines.

9 Machine Tool Improvement through Reverse Engineering and Computational Analysis with an Emphasis on Sustainable Design

Kaan Buyuktas, Muhammad Suhaib, Waqar Joyia, Kemran Karimov, Ihtesham Khan, and Volkan Esat
Middle East Technical University

CONTENTS

9.1 INTRODUCTION

The pertinence of reducing time for the development and production cycles has recently increased emanating from the escalation of globalization, hence resulting in heightened competition between companies. However, shortening the development cycle of a product requires radical amelioration of the current manufacturing techniques and is not easy. Therefore, the concerned companies are investing heavily in Reverse Engineering techniques to aid them in achieving reduced production cycles.

Reverse Engineering is basically the creation of engineering design data from already manufactured parts [1]. This differs it from the classical manufacturing process which starts with the design of the computer-aided design (CAD) solid model and ends with the actual manufactured product [2]. For reverse engineering, it is the stark opposite as the process begins with the product itself and ends up with its CAD model. In the past, reverse engineering was considered analogous to replication and was a violation of copyright patent law. This was since, in the past, reverse engineering was purely done from a geometric perspective which curtailed the retrieval of the intrinsic knowledge embedded in the functionality of the product itself [3]. Due to this very reason, the concept of functional reverse engineering was introduced to the manufacturing industry [4]. Functional reverse engineering entails the retrieval of the design intent of the product so that its working methodology can be comprehended. This allows for the exploration of superior design ideas with better efficiencies and better designs allowing for the legal application of reverse engineering to produce variations of old products or new products entirely in a short span of time. It is done through the bidirectional data transfer between the physical and digital form of the product [3].

Nowadays, reverse engineering is being implemented in various industries. Their usage covers a wide spectrum. Most commonly, it is done when retrieval of design documentation of a product is not possible or when the product is no longer in production. However, it can also be utilized to improve product performance and features by exploring and experimenting with design features, strengthening features of a product to elongate its longevity, and eliminating bad features. Reverse engineering also has a cross-industry appeal as its relevance is also present in other industries which are not related to manufacturing. It is used to generate data for the creation of surgical prosthetics and artificially engineered body parts, obtain 3-D data from an object or a person to be used in games and movies, and obtain architectural documentation [5].

The main aim of this chapter is to explore a redesign methodology by applying the fundamental laws of both geometric reverse engineering and functional reverse engineering to two different types of machine tools: a conventional drilling machine and a band saw. The focus will be on the processes involved in the manufacturing of the machine tools which are then amalgamated with certain extensions of contemporary techniques of engineering design to serve the utilitarian purpose of reverse engineering. Several new techniques currently being introduced in the field of reverse engineering are also explicitly elucidated. The redesigned critical parts of the machine tools are then subjected to finite element analyses (FEA) to get an assessment of mechanical loading on the critical parts of the machines, followed by environmental sustainability and cost analyses in order to assess the overall feasibility of the improved machine tools.

9.2 REVERSE ENGINEERING

Engineering entails the designing and manufacturing of a plethora of varying products. This can be done using two methodologies: forward engineering and reverse engineering. Forward engineering is the commonly used methodology which has been in existence for a long time. It requires the realization of an abstract ideology into a physical model by the mere use of logical engineering designing. In contrast, reverse engineering is the stark opposite of this as it initiates with the existing physical model and ends up with its duplication.

Reverse engineering is the process of obtaining CAD models for a physical product. Yau [6] eloquently described reverse engineering as the process of retrieving the geometric characteristics of a manufactured product by its digitization. It is achieved through the procurement of the minutiae of engineering design by evaluating the geometric dimensions and/or the functional model of the product. In the past few years, the relevance of reverse engineering has surged due to its cross-industrial appeal and is now considered to be a vital part of the product design and development cycle. Reverse engineering is being used for several purposes such as acquiring design data for a product whose design documentation has been lost and is no longer being produced, to analyze the design of a product to strengthen its good features and lessen its bad features to guarantee a more efficient functioning, obtaining data for the creation of dental and surgical prosthetics and artificially engineered body parts, acquiring 3D data from an object to be used for movies or games, and to generate architectural data [7,8]. This list is by no means exhaustive and with the progression of time several inventive usages of reverse engineering are being introduced.

9.2.1 Geometric Reverse Engineering

Geometric reverse engineering is the basic form of reverse engineering in which an existing part or product is converted into a 3D model so that a one-to-one clone of that specific part can be produced. It is sometimes considered to be analogous to duplication; however, that is a generalization because duplication is performed to make profit of an existing product whereas geometric reverse engineering has broader implications as it is based on long-term benefits and innovations [8]. Despite the differences, geometric reverse engineering is being used to produce replicas of certain products with less functionality.

In industry, geometric reverse engineering is usually done when there is a need to manufacture a damaged part which is no longer in production or to obtain technical drawing for old parts which do not have any technical data available [9].

9.2.2 Functional Reverse Engineering

Functional reverse engineering is a branch of reverse engineering which focuses on tacit knowledge that goes into designing a product rather than just focusing on the geometric structure. It can be defined as the process of unveiling the functionality of a product so that it is possible to make certain improvements to the existing model. This method of reverse engineering has recently gained traction as opposed to simple geometric reverse engineering as it does not carry the risk of copyright infringement of patent laws whereas just copying the original design could potentially lead to such an issue.

When materializing an original product, the designer has to go through three development phases which can be generalized as the conceptual design phase, the actual design phase, and the manufacturing phase. The designer's creativity for the product design is at its peak when in the conceptual design phase, where the designer assesses the probable solutions to the product function. As the phases progress, the creativity decreases because the conception starts to take a physical form. The actual design phase involves the realization of the geometric design which then leads to its production in the manufacturing phase. It is apparent that design intent and real design knowledge originate in the conceptual design phase and converge in the real design phase. Ergo, functional reverse engineering not only aids us in figuring out the ways in which the product works but also helps in ruling out the nonworking ways.

The function of a proposed product is arguably the most crucial factor in product design. Conventionally, the stipulations for a product are elucidated in the beginning so that relevant solution principles can be figured out. As Shimomura et al. also states that one of the key aims of functional reverse engineering is to unravel the function modelling so that an insight into the designer's intent can be visualized [10]. This is further supported by Fantoni et al. [11] in his research where he states that an appropriate function representation through function modelling for a product allows for meaningful methodology which can be reused to obtain results. Since then, function modelling has been used to obtain a set of sub-functions which can be extensively used to deduce overall functions. However, it is an arduous task to come up with optimum functions for products of varying degrees. In order to realize this approach, Cross proposed a technique way back in 2002 which involved the assessment of unknown constraints and criteria on the basis of the solution to the function so that the function can be reformulated with the inconsistencies accounted for [12].

Previously, the focus of researchers has been on the "function to form" approach. However, some researchers have investigated the "form to function" approach and as a result have devised a generalized approach for case-based functional design to redesign existing products with function predication [13,14]. Henceforth, due to the efforts of multiple researchers, a general catalog of common vocabularies and functions is available which will aid in the development of models by linking existing solutions to connecting functions [15].

Functional Reverse engineering is mainly carried out in three phases. For the first phase, the nature of the product to be reverse engineered needs to be determined. It is also referred to as the "prescreening phase". Next, information pertaining to the working of the product needs to be discerned so that the physical model can be converted into a functional model. Lastly, the functional model will be altered to recreate a new product with the same purpose.

9.2.3 Stages of Reverse Engineering

In the past, the process for reverse engineering had been quite straightforward. Conventional measuring tools such as micrometer screw gage, Vernier caliper, measuring tape, and inclinometer were used to get the general dimensions for the product which is to be reverse engineered. The measurements taken using these tools were then utilized to generate CAD models using software such as AutoDesk Inventor or SolidWorks. The 3D geometric model would then be used to either update the design with better design features leading to functional reverse engineering or clone the product. The conventional reverse engineering method has had a lot of flaws as potentially incorporating plenty of human measuring errors occurring during the measuring phase, resulting in inconsistent results. Due to this very reason the industry used to see reverse engineering with a cautious eye considering it to be more of a liability.

However, research works focused on reverse engineering have enabled it to improve with the progression of time. Nowadays, reverse engineering is performed using various software allowing for a high level of consistency. The modern reverse engineering method is divided into three phases. The first phase is known as scanning in which contact and/or noncontact scanning devices are used to map the geometrical features of the product into a point cloud form [16]. The next phase then makes use of a software which processes these clouds of points to remove errors and finely tune the data which can then be exported to CAD modelling. The second and third phases are called the point processing phase, and the geometric model development phase, respectively. This method can be completely automated provided that the equipment is available.

Once the object for reverse engineering has been selected, the scanning phase starts. It is done using multiple types of scanners to capture data pertaining to all the geometric features of the product. Scanners are divided into two main types which are contact scanners and noncontact scanners. In the beginning the size and shape of the part being reverse engineered need to be considered. This is done so that the type of measuring devices can be decided. Contact or noncontact scanners can be used to scan the part. This is done to capture the general geometric dimensions of the part. Contact scanners consist of various probes which track the contours of the surface so that the overall geometry can be mapped. Contact scanning devices are usually quite accurate with a tolerance range of about 0.01–0.02 mm. The drawback of contact method is that it is generally slow as it has to trace each and every point on the surface of the part meaning that big products would prove to be troublesome. Also, they require a bit of pressure to read the surface accurately meaning that soft materials cannot be scanned using this method [17].

Noncontact scanners are relatively new and make use of lasers and optics to scan the part without any physical contact. This allows for scanning of large parts at a much faster pace; however, there are a few drawbacks due to which they appear to be liabilities sometimes. Their tolerance ranges between 0.025 and 0.2 mm making them quite inconsistent for an inexperienced user. Also, laser-type scanners use light which bounces off from products with shiny surfaces, resulting in errors [18–20]. After accumulating the data from the scanners, the point cloud is edited to reduce noise and filter out the errors so that they can be transferred to a CAD modelling software for the development of 3D CAD models, which will then be used to physically produce the part.

The modern contact and noncontact scanning devices provide more convenient solution for obtaining the dimensions of a product to produce 3D CAD models, when compared to conventional measuring techniques; noncontact scanning takes relatively less time and the results obtained from the scanning process can be more reliable if the operator is more experienced [8]. But the scanning devices are quite expensive and the whole process of scanning a product and then analyzing the data collected requires high expertise, especially when processing the point cloud data, which is a complicated process. On the other hand, conventional measuring techniques, such as the use of calipers for measuring dimensions, offer an affordable solution with less expertise required, as it can be performed by anyone who is experienced with the basic measuring devices. In this chapter, the two case studies that are provided in the following sections are based on conventional measuring techniques, since the contact and noncontact scanning technology was not available at the venue and the budget and time limitation did not allow for new technology investment for the case studies.

9.3 CASE STUDY I: IMPROVEMENT OF A BENCH-TYPE DRILLING MACHINE

In the first case study, to demonstrate the proposed reverse engineering platform, a bench-type drilling machine was chosen, which was located at the machine shop facility of METU NCC. Drilling machines are one of the most basic machine tools in any machine shop, which can be used for machining holes through several materials such as metals or wood. They contain several parts such as pulleys, belts, motors, quills or sleeves, spindles, columns, heads, bases, drill chucks, and worktables. The bench-type drilling machine chosen for this case is presented in Figure 9.1 below.

First, the drilling machine was thoroughly examined to identify all the parts and subassemblies that constitute the machine tool. Then, the drilling machine was disassembled into subassemblies, and then into individual parts. Next, measurements were taken of all the parts using several measuring instruments. The parts were created and assembled in Autodesk Inventor Professional 2018. Since the objective was to improve the drilling machine, the most critical component was to be identified in the assembly. FEA were to be performed on that part to determine the most critical regions on it so that a stronger product with a higher factor of safety could be proposed. For this machine tool, the identified part turned out to be the spindle.

FIGURE 9.1 Bench-type drilling machine located at METU NCC.

Spindle is generally considered to be the heart of a drilling machine. In addition to the original model, two new models for the spindle were created in the software medium with different geometries. FEA was performed on all of them. This was followed by the strength, sustainability, and cost analysis of all the spindle models with different materials, using an evaluation matrix, in order to determine the most feasible case which best suited the design and safety criteria for the drilling machine.

The whole process of applying reverse engineering on the drilling machine was divided into several stages, which are illustrated in the flow chart in Figure 9.2.

9.3.1 Disassembly of the Drilling Machine

Initially, the main assembly was disassembled into subassemblies as shown in Figure 9.3 below. The subassemblies were further disassembled into individual components so all the necessary dimensions could be taken. Different types of wrenches were used in the disassembling process.

Stage 1
Dissassembly of the drilling machine

Stage 2
Measurement process of all the parts that constitute the drilling machine

Stage 3
3D CAD modelling of the drilling machine with the original dimensions

Stage 4
General analysis of the selected crucial component (spindle) of the drilling machine

Stage 5
Finite Element Analysis (FEA) of the spindle

Stage 6
Designing new geometrical models of the spindle for improvements

Stage 7
Eco-material / sustainibility and cost analysis of the spindle models

Stage 8
Utilizing evaluation matrix for choosing the most feasible spindle model

FIGURE 9.2 Flow chart for the reverse engineering of the drilling machine.

FIGURE 9.3 Disassembly process.

9.3.2 Measurement Techniques

Conventional measurement instruments were used for measuring the dimensions of the parts or subassemblies. These instruments included Vernier calipers, digital Vernier calipers, measuring tapes, and inclinometers. These measurement instruments are provided in Figure 9.4 below.

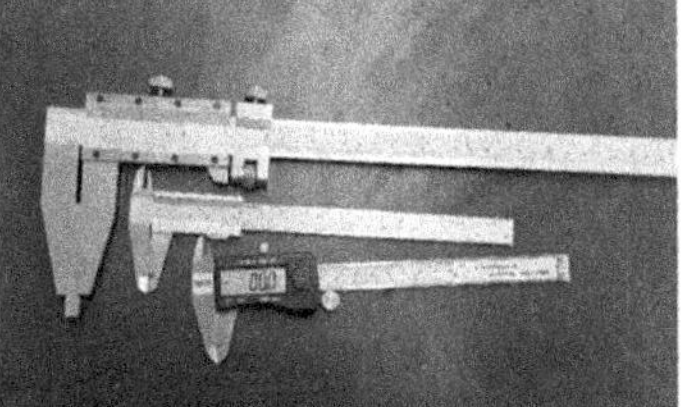

FIGURE 9.4 Tape measure, Vernier calipers, and inclinometer.

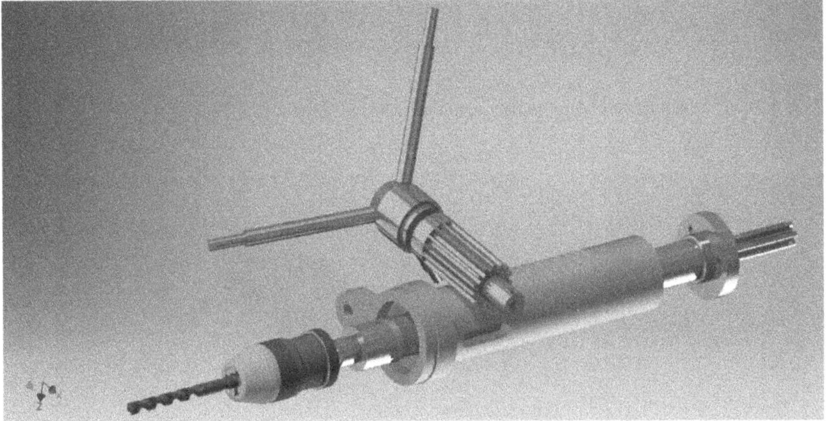

FIGURE 9.5 Assembly of the quill mechanism.

This stage was the most time-consuming one since there were many dimensions and each dimension was taken with utmost care to avoid errors. For instance, in the pulley mechanism, threads were used to measure the internal diameters. Also, the digital Vernier caliper gave better results than the regular Vernier caliper; therefore, it was used for most of the dimensions.

9.3.3 CAD Modelling

In the next stage, Autodesk Inventor Professional 2018 was used to create the solid models in 3D. Initially, the main parts for each mechanism were modelled in the software. Then, the created 3D parts were assembled together, and the final assembly was obtained using the necessary constraints. The assembly for the quill mechanism is shown in Figure 9.5.

As clearly visible in Figure 9.5, the spindle holds the drill bit with the help of the drill chuck. The sleeve or quill holds the spindle, which revolves around the fixed axis concentric with the axis of the sleeve. The assembly is held vertically in the drilling machine.

Another important assembly is the one for the pulley and belt mechanism, provided in Figure 9.6. In this assembly, there are two pulleys. The black one is connected to the spindle, whereas the silver one is fixed on the motor. The motor provides

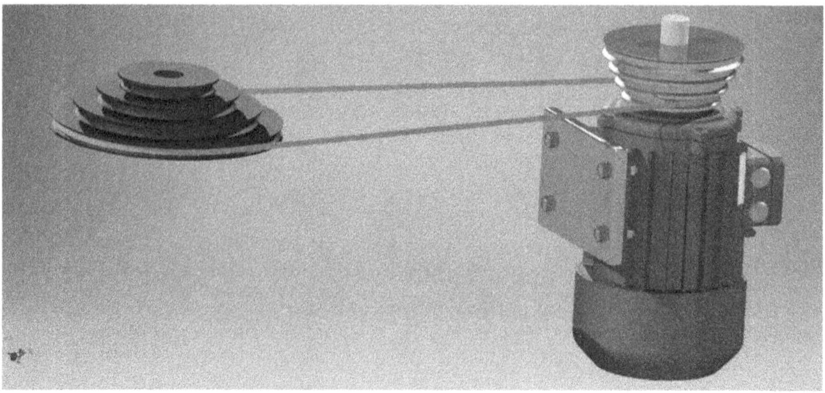

FIGURE 9.6 Assembly of the pulley and belt mechanism.

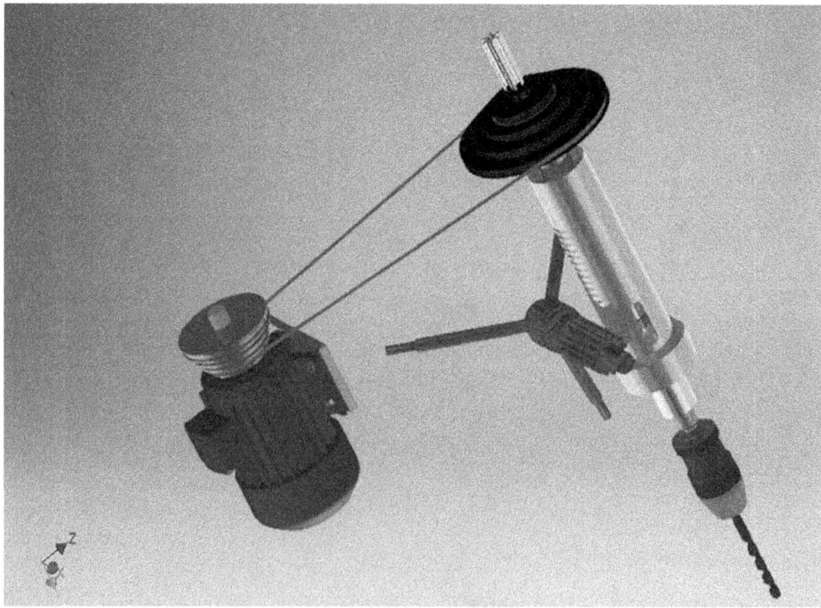

FIGURE 9.7 Interior assembly.

power to the spindle with the help of this mechanism. Changing the gear ratios on the pulleys changes the torque and the rotational speed of the spindle.

The interior assembly of the drilling machine is shown in Figure 9.7 below. It consists of both the quill mechanism, and the pulley and belt mechanism.

The exterior casing containing the interior assembly is given in Figure 9.8a, whereas the complete assembly of the Bench-type drilling machine is shown in Figure 9.8b. The solid model assembly comprises all the subassemblies as well as the parts such as nuts and bolts and the bearings.

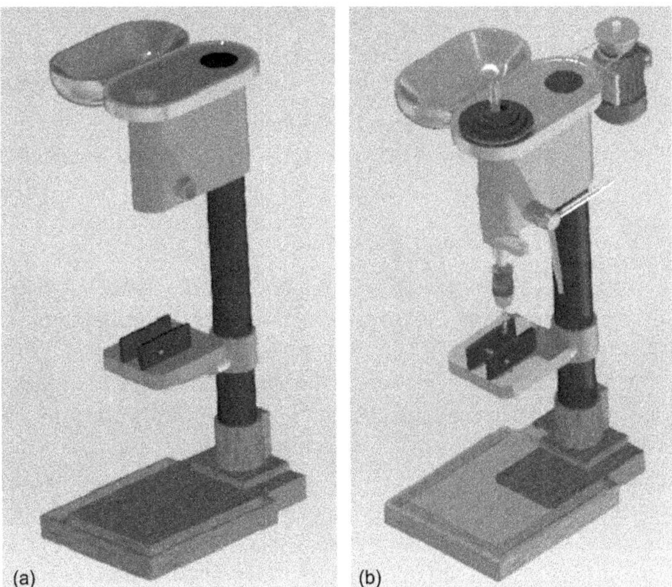

FIGURE 9.8 (a) Exterior casing assembly, and (b) assembly of the entire bench-type drilling machine.

9.3.4 Dynamic Analysis of the Selected Component of the Drilling Machine

The spindle is arguably the most critical component of the drilling machine since the major stresses occur within the spindle when the motor power is transmitted to the drill bit through the spindle. Due its long length and geometry, it experiences large stresses during operation. Also, different drilling machines have different configurations for their spindle meaning that the spindle is the part which makes each drill machine unique. The spindle in the actual drilling machine and its designed CAD model are shown below in Figure 9.9.

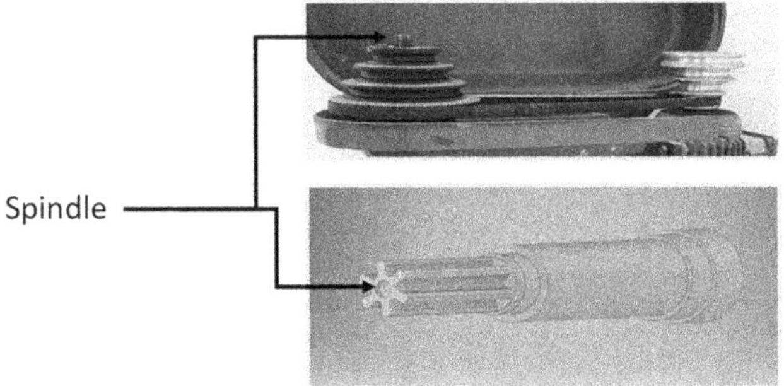

FIGURE 9.9 Spindle: heart of the drilling machine.

TABLE 9.1
Specifications of the Motor Connected to the Pulley and Belt Mechanism

Motor Power, P (kW)	Rotational Speed of the Pulley Connected to Motor, ω_2 (rpm)	Motor Power Efficiency, η (%)
0.55	1365	80

Since the spindle is the most crucial component in the assembly, it is important to determine the torque acting on it, resulting from the transmission of power from the motor through the pulley and belt mechanism. The torque value can be used to obtain the Von Mises stress distribution on the spindle using the FEA technique. The specifications of the motor used in the assembly of the drilling machine are provided in Table 9.1 below.

The torque acting on the spindle must be maximized to obtain the maximized Von Mises stress distribution for the spindle. Since torque is inversely proportional to rotational speed, minimizing the rotational speed will maximize the torque. The speed ratio of the pulley was used to minimize the rotational speed of the pulley connected to the spindle and thus the rotational speed of the spindle. From Equation (9.1) and Figure 9.10, it can be deduced that the step on Pulley 1 that gives the maximum radius will give the minimum rotational speed and thus the maximum torque. The adjusted configuration for the pulley and belt mechanism, which results in maximum torque, is shown in Figure 9.10.

$$\frac{\omega_1}{\omega_2} = \frac{r_2}{r_1} \tag{9.1}$$

The radius of the step of Pulley 1 was measured to be 82.76 mm, whereas the radius for the step of the other pulley was found to be 21.55 mm through the measurement process.

The equation for the speed ratio in Equation (9.1) was used to calculate the rotational speed of Pulley 1. ω_1 was found to be 355.7 rpm. Using the efficiency as 80%, the power P_1 transmitted to the pulley was found to be 440 W, using Equation (9.2).

$$P_1 = \eta P \tag{9.2}$$

Finally, ω_1 and P_1 were used to find the torque T on the spindle using Equation (9.3).

$$T = \frac{P}{\omega} \tag{9.3}$$

The value of the torque on the spindle was found as 11.82 Nm.

FIGURE 9.10 Pulleys and belt.

9.3.5 FEA of the Spindle

To make the spindle stronger through increasing its factor of safety, FEA of the spindle was carried out first. FEA was utilized to determine the most critical regions on the spindle where the stresses are maximum as those regions can then be modified to propose an enhanced product. Autodesk Inventor Professional 2018 provides stress analysis module; however, in order to provide highly reliable results, a powerful FEA software was preferred. Therefore, in this case study, the FEA of the spindle was performed using Autodesk Nastran In-CAD 2018, which is directly embedded in Autodesk Inventor Professional software, which can potentially provide highly realistic and satisfactory results (comparable to those from ANSYS, Marc, or ABAQUS). Nonlinear static analysis was used for the FEA of the spindle.

The first step in FEA technique was to conduct sensitivity analysis to observe how the maximum von Mises stress varies with increasing number of the elements. Theoretically, von Mises stress values are expected to converge at a certain point with increasing number of elements. When the number of elements was increased from 3270 to 5326, the maximum von Mises stress fluctuates within the range 22–24 MPa. With the further increase in the number of elements, the maximum von Mises stress showed a sharp rise, reaching a value of 46.9 MPa when the number of elements was 22,332. However, the value for the maximum von Mises Stress started to converge when the number of elements was raised further by large numbers, after this point. It reached a maximum value of 58.7 MPa with the mesh size of 0.002 mm and the number of elements was 245,069. Maximum Von Mises stress began to converge around this point, which provided the minimum number of elements for the results not to be affected from the mesh size. The number of elements was not increased further as the differences in the results were insignificant, and also the solver stopped working when the number of elements reaches beyond a specific number due to the limitation of the computational power of the computer. The results are depicted in Figure 9.11.

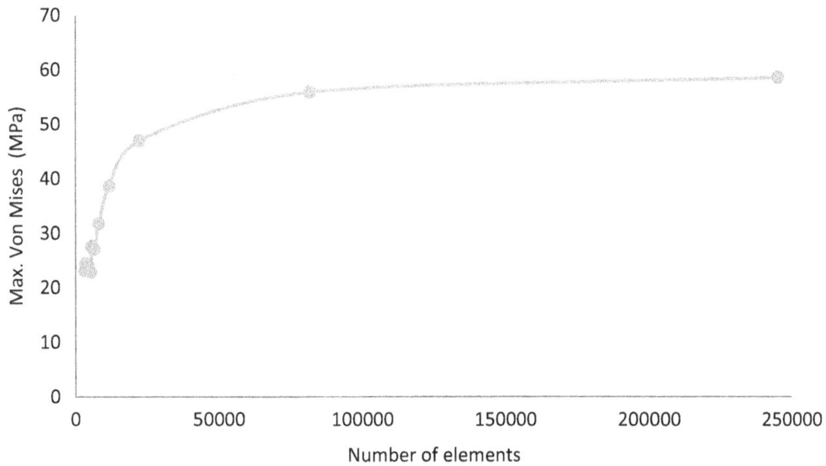

FIGURE 9.11 Maximum von Mises stress vs number of elements.

Figure 9.11 shows the steep line at the beginning of the graph which is due to inadequate mesh size. The visual simulations for these different mesh sizes can be seen in Figure 9.12 along with the most critical section at the root of the spindle keyways.

As the critical region was identified on the spindle, the next step was to figure out an alternate approach that could help in finding the exact value of the maximum Von Mises stress. In the next stage of the sensitivity analysis, only the mesh size for

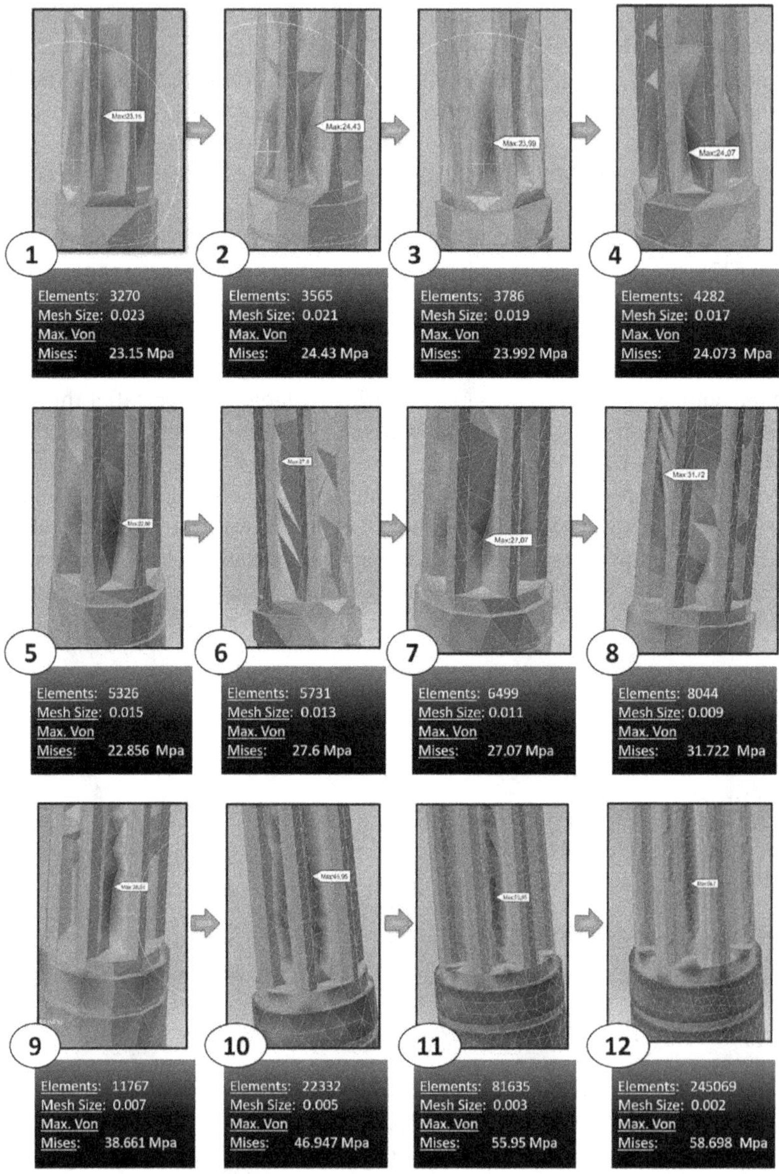

FIGURE 9.12 FEA results during sensitivity analysis.

the critical regions was decreased while the general mesh size remained constant with a higher value than that of the critical region; in other words, the variations in von Mises stress were observed by increasing the number of elements only in the critical zone of the spindle, while holding the general mesh size constant. With this approach, higher accuracy was achieved with fewer number of elements. The resulting mesh for the new approach is shown in Figure 9.13 below.

In Figure 9.13, the grey arrow represents the torque applied on the spindle in the upper region which is indirectly in contact with the pulley. The bottom surface of the spindle is fixed, shown by the white arrows. Furthermore, the mesh is denser in the upper (critical) region and coarser in the lower region. The results are illustrated in Figure 9.14. The maximum Von Mises stresses not only converged but almost stopped changing when the number of elements was increased from 129,574 to 270,117. This indicates that this converged value is quite accurate. Moreover, the converged value obtained from this analysis (61.7 MPa) is quite close to the one obtained from the previous approach (58.6 MPa), further ensuring the accuracy of the alternate approach. Hence, the alternate approach, with the critical zone mesh size of 0.001 and the overall mesh size of 0.01, was decided to be used for all of the oncoming analyses of the spindle.

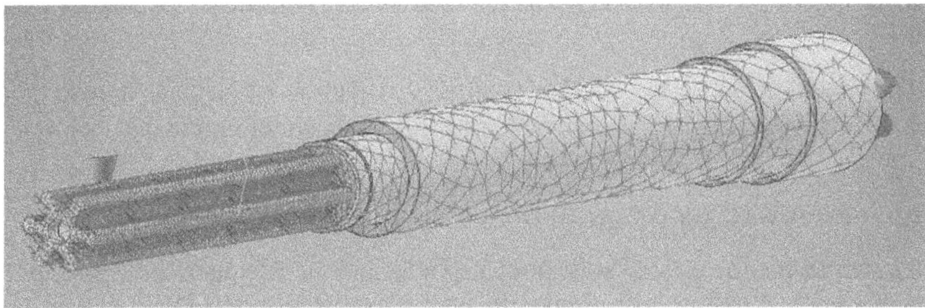

FIGURE 9.13 Meshed regions on the spindle.

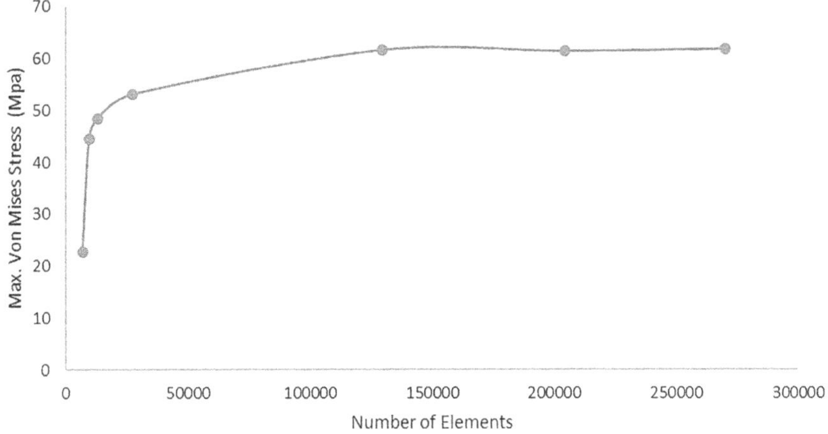

FIGURE 9.14 Maximum von Mises stress vs number of elements for the second approach

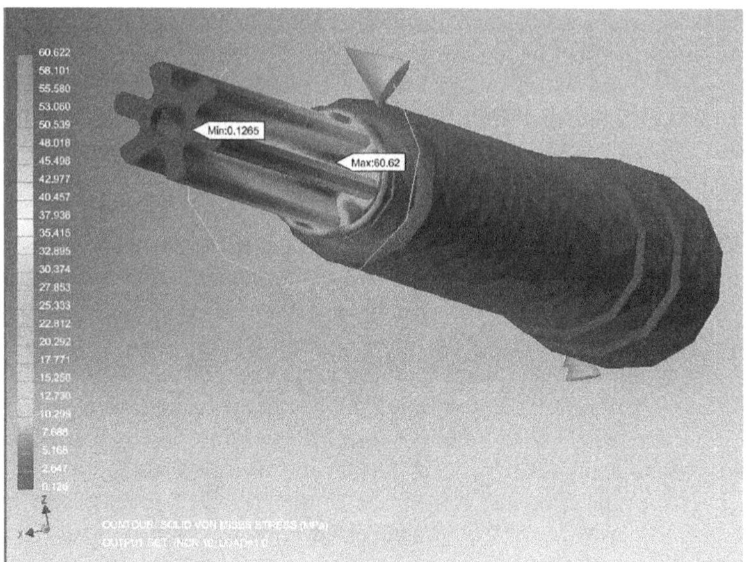

FIGURE 9.15 FEA visualization for the von Mises stress distribution of the original spindle model.

The visualization for the Von Mises stress distribution for original spindle model with the material assigned as malleable cast iron is shown in Figure 9.15 below.

9.3.6 Improving the Strength of the Spindle: Designing New Models

After identifying the critical zone on the spindle and the adequate mesh size for the critical zone and the overall spindle, the next step was to modify the geometries of the critical regions and see how different geometries affect the maximum von Mises stresses in the spindle. The critical zone was basically the region between the keyway fins at the root of the spindle. Therefore, two additional spindle models were designed in Autodesk Inventor Professional 2018, based on the previous FEA analyses, with different number of fins. The new proposed models are shown in Figure 9.16 below.

As it can be seen in Figure 9.16, the actual model contained 6 fins. The first newly designed spindle model had 5 fins, whereas the second one had 7 fins, as shown with the arrows in Figure 9.16. FEA was carried on both new models using the previously determined mesh size values (0.001 for critical zone & 0.01 for overall spindle). The same software and process of FEA (which was used to analyze the original model) were used for the new models. Furthermore, a total of seven cases were used for the strength, sustainability, and cost analysis of the spindle models, explained thoroughly in the following sections.

9.3.7 Strength Analysis of the Spindle Models

After creating the 3D spindle solid models, a total of seven different cases were chosen for analysis. For the original model and the model with 5 fins, three different

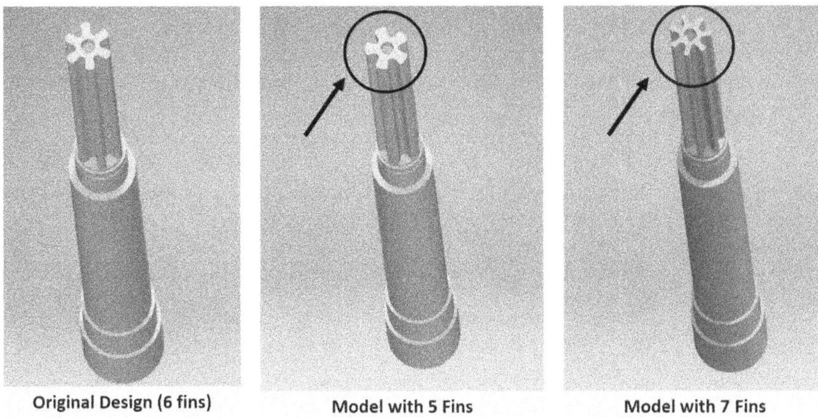

FIGURE 9.16 3D CAD solid models of the spindle.

materials were assigned to each one of them; the materials were Aluminum 6061, Stainless Steel AISI 440C, and Malleable Cast Iron. The original 3D spindle model which reflects the actual spindle from the drilling machine was assigned Malleable Cast Iron material, shown by the highlighted region in Analysis 6 below in Table 9.2. Also, only one case was chosen for the spindle model with 7 fins, since the stresses developed in that model were the maximum compared to the rest; choosing more material cases for the model with 7 fins would not benefit the analysis much. The material properties along with the maximum von Mises stress and the factor of safety for all of the seven cases are listed below in Table 9.2 for comparison. The factor of safety was found using Equation (9.4) below.

$$\text{Factor of safety} = \frac{\text{Tensile Yeild Strength}}{\text{Max. Von Mises Stress}} \tag{9.4}$$

From Table 9.2, it can be deduced that the spindle model with 5 fins and Stainless Steel AISI 440C material, represented by the green region or Analysis 2, has the highest factor of safety (about 10.1), leading in strength analysis; one reason for this case being the safest option in terms of strength is that the number of fins is reduced and the material chosen has the highest yield strength, giving the largest factor of safety compared to the other cases. However, the maximum von Mises stress is minimum in analysis 1, but the yield strength of material is low compared to Stainless Steel AISI 440C. The original spindle model, shown by Analysis 6, shows very low factor of safety compared to Analysis 2. Hence, the spindle in Case 2 can be chosen for the drilling machine if strength is the main criterion for design.

9.3.8 Sustainability and Cost Analyses of the Spindle Models

The ECO Material Advisor feature of Autodesk Inventor Professional 2018 was used to determine the environmental impacts and the cost associated with the materials chosen for the spindle models. The key environmental indicators were estimated

TABLE 9.2
Material Properties, Maximum Von Mises Stress, and Factor of Safeties for the Spindles

Analysis No. / Case	Number of Elements	Spindle Geometry	Material	*E* (GPa)	Poison's Ratio	Tensile Yield Strength (MPa)	Max. Von Mises Stress (MPa)	Factor of Safety (F.S.)
1	120005	5 Fins	Aluminum 6061	68.9	0.33	276	49.320	5.60
2	120005	5 Fins	Stainless Steel AISI 440C	206.7	0.28	500	49.504	**10.10**
3	120005	5 Fins	Malleable Cast Iron	165	0.26	205	49.574	4.14
4	128842	6 Fins	Aluminum 6061	68.9	0.33	276	60.606	4.55
5	128842	6 Fins	Stainless Steel AISI 440C	206.7	0.28	500	60.618	8.25
6	128842	6 Fins	Malleable Cast Iron	165	0.26	205	60.622	3.38
7	157269	7 Fins	Malleable Cast Iron	165	0.26	205	64.077	3.20

using this feature, including the quantity of carbon dioxide involved in the creation or disposal of the spindle, the energy used in the manufacturing stage and disposal of the product, and the estimated cost of the materials assigned to the spindle models. The contribution of the manufacturing techniques was also considered in the sustainability analysis. The detailed results for all of the seven cases are provided in Table 9.3 below.

The original spindle mode is represented by Case 6 as highlighted, having the carbon dioxide footprint as 2.74 kg, energy usage of 36.1 MJ in the manufacturing and disposal process, and the cost of material involved as 1.16 US Dollars. The manufacturing technique chosen was casting, since the drilling machine is quite old, and the actual spindle was most probably manufactured using casting process. The values for all the mentioned environmental indicators are minimum in Case 7 (green region in Table 9.3), when the spindle model with 7 fins is chosen; thus, Case 7 provides the most sustainable option. There is a trade-off between the strength of the materials and the sustainability of the product. If the highest concern is the sustainability of the machine element, then Case 7 provides the best option. It should be noted that there is about a 20% difference between the CO_2 footprint values for Case 7 and Case 2; however, since this spindle is not to be mass produced and the primary objective is to provide a stronger spindle in hopes of providing a safer working environment for students who are usually inexperienced and can potentially misuse the machine, Case 2 can still be considered. This will result in a higher carbon dioxide footprint, but the factor of safety will increase by a lot.

TABLE 9.3
Results for Sustainability and Cost Analyses of Different Cases of the Spindle Model

Analysis No. / Case	Spindle Geometry	Material	CO_2 Footprint (kg)	Energy Usage (MJ)	(Material) Cost (USD)	Manufacturing Technique
1	5 Fins	Aluminum 6061	3.32	42.0	2.15	Metal Powder Forming
2	5 Fins	Stainless Steel AISI 440C	3.30	42.7	3.25	Forging/Rolling
3	5 Fins	Cast Iron, Malleable	2.79	36.8	1.18	Casting
4	6 Fins	Aluminum 6061	3.26	41.3	2.11	Metal Powder Forming
5	6 Fins	Stainless Steel AISI 440C	3.24	41.9	3.19	Forging/Rolling
6	6 Fins	Cast Iron, Malleable	2.74	36.1	1.16	Casting
7	7 Fins	Cast Iron, Malleable	**2.70**	**35.6**	**1.14**	Casting

9.3.9 Evaluation Matrix Approach for Selecting the Most Feasible Spindle Model

After carrying out the strength, sustainability, and cost analyses for all the cases, the results were evaluated to determine the most feasible spindle model for the drilling machine. A scale of 1–7 was chosen for evaluation of each of the factor involved in the analysis, where 1 represents the least preferred option and 7 indicates the most preferred choice. Then, to each of the factor in the analysis, weights were assigned based on the level of importance and were incorporated in points for the scale chosen. The assigned weights include Factor of Safety (70%), CO_2 Footprint (5%), Energy Usage (15%), and Material Cost (10%). The weightage for CO_2 footprint and energy usage was kept low because the spindle was not to be mass produced and for the scope of the study, the improvement of strength of the spindle was of more importance. If the spindle was to be mass produced, then the CO_2 footprint and energy usage would carry much more weightage as their impact would be higher. The results are presented in the evaluation matrix in Table 9.4 below.

As shown in Table 9.4, based on the points assigned to each factor, the actual model received 3.20 (weighted) ranking points (highlighted by tan region), which is considerably low when compared to the other cases, since the factor of safety was very small for the actual model compared to others. On the other hand, Case 2 (green region) with the spindle model of 5 fins and stainless steel AISI 440C material has the highest (weighted) ranking points with the value of 5.25. Therefore, the most preferred option based on the assigned weights to the factor of safety, environmental factors, and the cost of the material was Case 2, which provides a stronger spindle product that can be used in the drilling machine thus providing a higher factor of safety.

TABLE 9.4
Evaluation Matrix for the Strength, Sustainability, and Cost Analysis of Different Cases of the Spindle Model

#	Spindle Geometry	Material	Factor of Safety (F.S.)	Points	CO_2 Footprint (kg)	Points	Energy Usage (MJ)	Points	(Material) Cost (USD)	Points	Manufacturing Technique	Total Points	Ranking Points (Weighted)
1	5 Fins	Aluminum 6061	5.596	5	3.32	1	42.0	2	2.15	3	Metal Powder Forming	11	4.15
2	5 Fins	Stainless Steel AISI 440C	10.100	7	3.30	2	42.7	1	3.25	1	Forging/Rolling	11	**5.25**
3	5 Fins	Malleable Cast Iron	4.135	3	2.79	5	36.8	5	1.18	5	Casting	18	3.60
4	6 Fins	Aluminum 6061	4.554	4	3.26	3	41.3	4	2.11	4	Metal Powder Forming	15	3.95
5	6 Fins	Stainless Steel AISI 440C	8.248	6	3.24	4	41.9	3	3.19	2	Forging/Rolling	15	5.05
6	6 Fins	Malleable Cast Iron	3.382	2	2.74	6	36.1	6	1.16	6	Casting	20	3.20
7	7 Fins	Malleable Cast Iron	3.199	1	2.70	7	35.6	7	1.14	7	Casting	22	2.80

9.4 CASE STUDY II: IMPROVEMENT OF A METAL/ WOOD CUTTING BAND SAW

As the second case study to further demonstrate the proposed reverse engineering platform, a metal/wood cutting band saw (JET VBS-18MW) was selected, also located at the machine shop facility of METU NCC. The machine tool can be seen in Figure 9.17. The band saw is vertically positioned with a triangular column and has two modes of operation as wood and metal modes. The band saw had the characteristic dimensions of 10-TPI metal cutting blade, approximately 3480 mm (L) × 19 mm (W) × 0.64 mm (Thick). Blade tension is adjusted with the handle that changes the compressive force for the spring, which is called as tension gauge in the manual of the saw. Electric motor equipped with a heavy-duty gear box allows the user to cut both wood and metal parts [21]. For speed of the band saw there are two set of

FIGURE 9.17 Metal/wood cutting band saw (JET VBS-18MW) and internal mechanism.

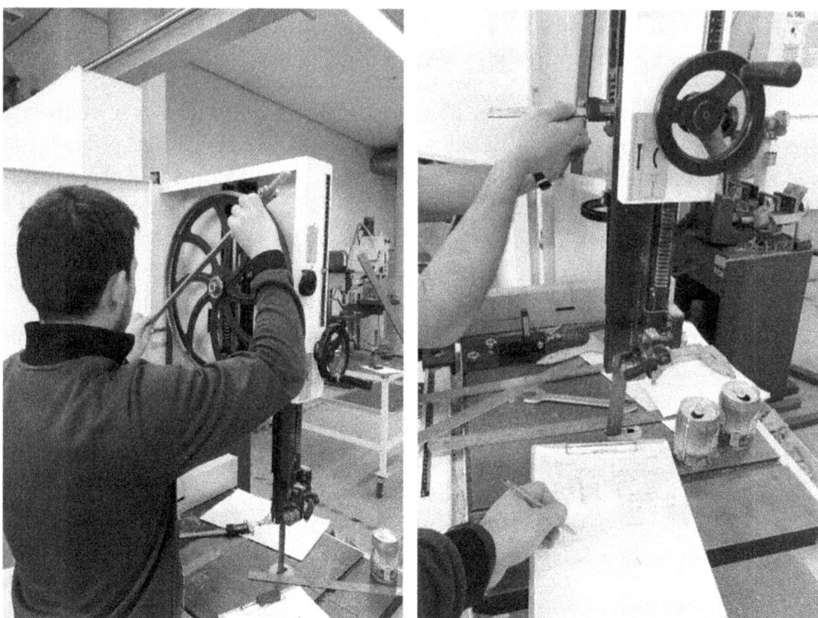

FIGURE 9.18 Collection of geometric data through measurements.

three belt positions, placed inside of the mechanism which is connected to bottom flywheel's shaft with a V-band. The measurements of the parts of the band saw were taken using common engineering measurement instruments as metric ruler, Vernier metric caliper/Vernier digital metric caliper, long jaw metric caliper, etc. Photos from the measurements and geometric data collection stage can be seen in Figure 9.18.

9.4.1 CAD Solid Modelling

Using the dimensions obtained from the measurements stage, the 3D CAD solid models of the parts were created one by one in Autodesk Inventor software. After part creation, parts were called back for the assembly of mechanism. The part list of the mechanism of the band saw is listed below:

• Upper Flywheel	• Upper Wheel Brackets
• Bottom Flywheel	• Spring
• 2×Plastic Cover	• Square Nut
• Upper Flywheel Pin	• Blade Adjusting Screw
• Ball Bearings	• Hand Wheel
• Band Saw	• Spindle Pulley
• Pin Bracket	• V-band
• Sliding Bracket	• Motor Pulley
• Nuts and Washers	

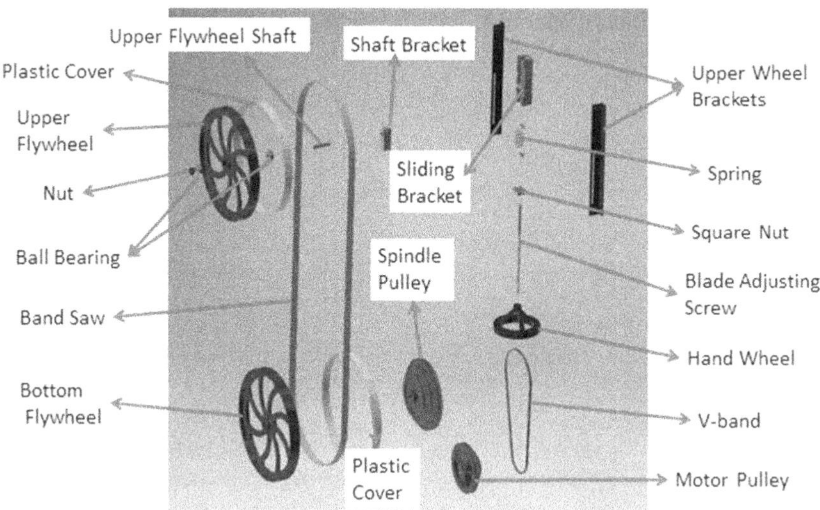

FIGURE 9.19 Exploded view of the band saw mechanism parts that form the overall assembly.

The main mechanism of the band saw assembly was segmented with the presentation tool of Autodesk Inventor for better observation and each part was named accordingly in Figure 9.19. Figure 9.20a and b show the picture designation method applied within Autodesk Inventor software. Some difficult regions of flywheel geometry could not be measured with the existing devices; therefore, photos were taken from a nonperspective view for complicated geometric features. Then, the photos were inserted into the software as a 2-D drawing model, scaled with known diameters of outer circle and inner circle. After scaling the image, the curves were obtained with maximum possible accuracy and centers of circular curves were fitted to the same point at a circle that was located at the center of the flywheel.

The assembly views of the developed solid model of the main band saw mechanism are given in Figure 9.21.

9.4.2 Analysis of the Band Saw Mechanism

Analysis of the band saw mechanism was carried out in Autodesk INVENTOR environment. Stress analysis gave results in terms of maximum von Misses stress distributions, maximum displacements, and factor of safeties. The most critical components of the mechanism were chosen to be analyzed – flywheel, band saw, and flywheel pin, i.e., the components that potentially experience the largest stresses and were prone to failure. Some design decisions taken are as follows:

- Two combinations of materials to be assigned are chosen for the mechanism components:
  1. Flywheel – Cast Iron (Gray), Band Saw – Stainless Steel (Austenitic), Flywheel Pin – Steel (Low Alloy)

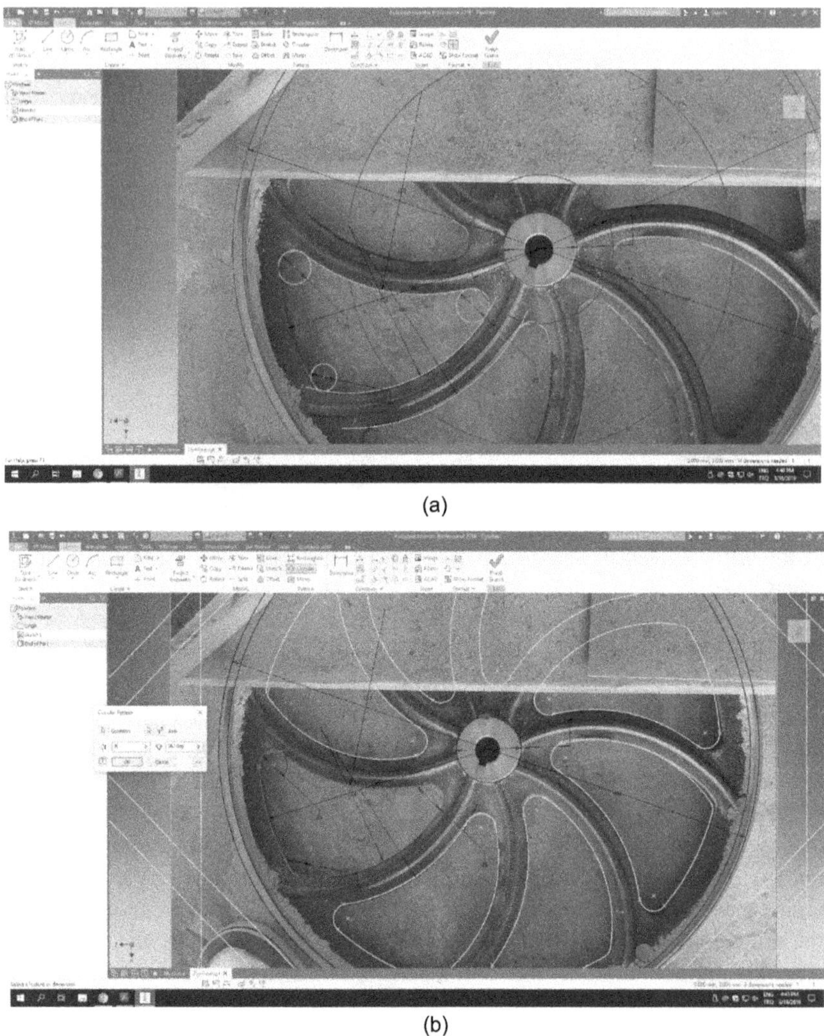

(a)

(b)

FIGURE 9.20 (a) and (b) Picture designation method applied within Autodesk Inventor software.

2. Flywheel – Steel (Cast), Band Saw – Steel (High Strength Low alloy), Flywheel Pin – Steel (High Strength Low Alloy)

- Only flywheel is subjected to geometrical modifications.
- Both material combinations were assigned to each geometrical modification of the flywheel such as Geometry 1: cast iron, Geometry 1: Steel, Geometry 2: Cast iron, Geometry 2: Steel, and so on.
- Yield strength instead of ultimate tensile strength (UTS) was selected to avoid plastic deformation, which makes product even safer.
- The back side of the flywheel pin (the side that is connected to the bracket) was attached to the vertical plane.

FIGURE 9.21 The assembly views of the developed solid model of the main band saw mechanism.

As shown in Figure 9.22, forces were set to the ends of the band saw. Both of these forces are tensile. Due to the friction force between the outer surfaces of the rim of the flywheel and surface of the belt, one of the forces must be larger than the other one. According to the belt design criterion, belting equation given in Equation (9.5) below describes the dependency of these forces, where F_1 and F_2 are the magnitudes of the tight and slack side forces, respectively, whereas φ is the belt wrap angle and μ is coefficient of friction.

$$F_1 = F_2 e^{\mu \varnothing} \tag{9.5}$$

Force 1 was figured out to be 3250 N, and Force 2 was computed to be 1500 N. Following the settings and preparation, finite element and sustainability analyses were performed.

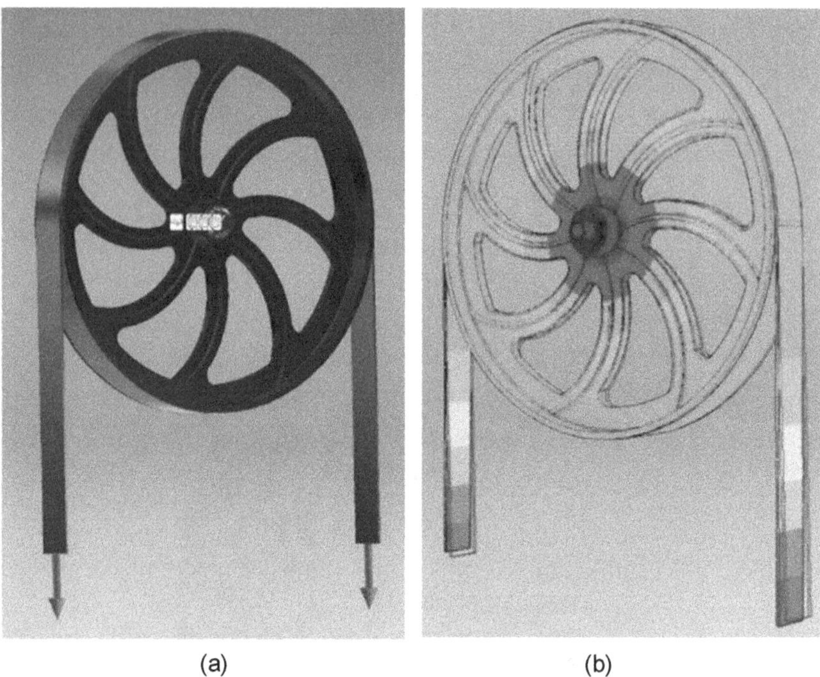

(a) (b)

FIGURE 9.22 (a) Flywheel and belt system. (b) FEA of the system.

9.4.3 FEA of the Flywheel

In the analysis of the flywheel, first design to be analyzed was the original actual design with material assigned as cast iron. Value of the maximum von Mises stress was found to be 138.8 MPa, mass was calculated as 7.578 kg, maximum displacement reached the point of 3.686 mm, and a minimum factor of safety turned out to be 5.30. After this, flywheel with the same design but with different material as steel (cast) was analyzed. The maximum von Mises stress reached the value of 176.5 MPa, the mass as 8.320 kg, maximum displacement observed turned out to be around 2.6 mm, and the minimum factor of safety is 5.12. Autodesk Inventor ECO Material Adviser analysis reveals the following output: energy required to produce such flywheel made of cast iron is approximately equal to 220 MJ, carbon dioxide emission is 14 kg, and water usage occurs to be 330 l; whereas for the original actual steel flywheel energy usage is 220 MJ, carbon dioxide footprint is 15 kg, and water used is around 380 l. Modifications were only conducted to the spokes of the flywheel. Spokes (arms) were decided to be thicker. The two materials mentioned above were assigned one by one and analyses were performed on each of them. Results are tabulated and presented in Figure 9.23. The proposed optimal modified design of the flywheel is shown in Figure 9.23, Image 3. The only modification is that it was subjected to be the simplified geometry of the spokes, which were designed straight. Such simplification also gave reasonably good results which are given in Figure 9.23. The last design is composed of simplified straight spoke and longer hub of the flywheel. All the factors of

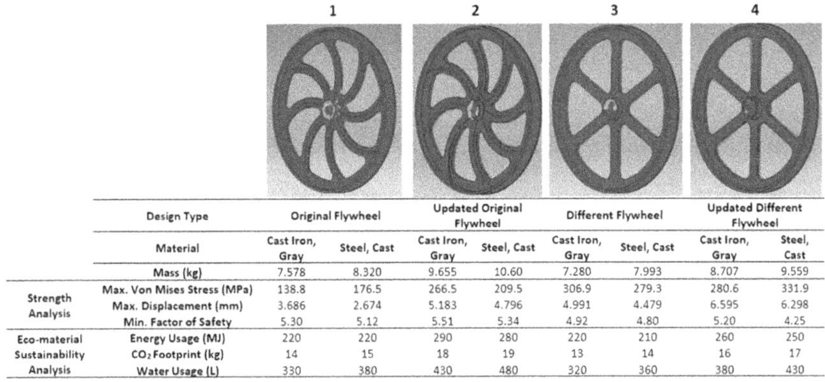

	Design Type	Original Flywheel		Updated Original Flywheel		Different Flywheel		Updated Different Flywheel	
	Material	Cast Iron, Gray	Steel, Cast	Cast Iron, Gray	Steel, Cast	Cast Iron, Gray	Steel, Cast	Cast Iron, Gray	Steel, Cast
	Mass (kg)	7.578	8.320	9.655	10.60	7.280	7.993	8.707	9.559
Strength Analysis	Max. Von Mises Stress (MPa)	138.8	176.5	266.5	209.5	306.9	279.3	280.6	331.9
	Max. Displacement (mm)	3.686	2.674	5.183	4.796	4.991	4.479	6.595	6.298
	Min. Factor of Safety	5.30	5.12	5.51	5.34	4.92	4.80	5.20	4.25
Eco-material Sustainability Analysis	Energy Usage (MJ)	220	220	290	280	220	210	260	250
	CO_2 Footprint (kg)	14	15	18	19	13	14	16	17
	Water Usage (L)	330	380	430	480	320	360	380	430

FIGURE 9.23 Flywheel designs and results of the FEA and sustainability analyses.

safety turned out to be very close to each other; therefore, selection criterion became production cost. The last design of the flywheel was made of gray cast iron as it is relatively easier to manufacture due to the geometry; therefore, production cost is expected to be much lower.

9.4.4 FEA of the Band Saw

Analysis of the band saw was simpler as it did not experience any changes and modification in its geometrical design. Analysis of the band saw was carried out to assess its response to the different flywheel designs as explained in Section 9.4.3. Figure 9.24 shows four images of the band saw's stress responses to different flywheel designs. Image 1 in Figure 9.24 shows the von Mises stress distribution along the band saw operating with originally designed flywheel. Images 2–4 in Figure 9.24 illustrate the von Mises stress distribution along the band saw at the flywheel with thicker spokes,

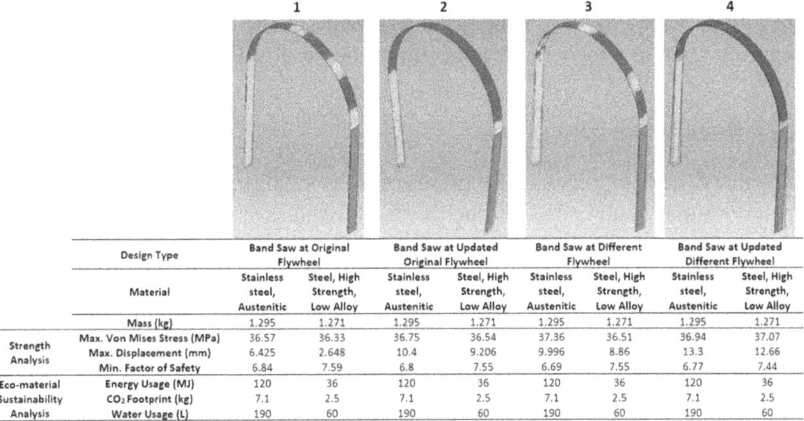

	Design Type	Band Saw at Original Flywheel		Band Saw at Updated Original Flywheel		Band Saw at Different Flywheel		Band Saw at Updated Different Flywheel	
	Material	Stainless steel, Austenitic	Steel, High Strength, Low Alloy	Stainless steel, Austenitic	Steel, High Strength, Low Alloy	Stainless steel, Austenitic	Steel, High Strength, Low Alloy	Stainless steel, Austenitic	Steel, High Strength, Low Alloy
	Mass (kg)	1.295	1.271	1.295	1.271	1.295	1.271	1.295	1.271
Strength Analysis	Max. Von Mises Stress (MPa)	36.57	36.33	36.75	36.54	37.36	36.51	36.94	37.07
	Max. Displacement (mm)	6.425	2.648	10.4	9.206	9.996	8.86	13.3	12.66
	Min. Factor of Safety	6.84	7.59	6.8	7.55	6.69	7.55	6.77	7.44
Eco-material Sustainability Analysis	Energy Usage (MJ)	120	36	120	36	120	36	120	36
	CO_2 Footprint (kg)	7.1	2.5	7.1	2.5	7.1	2.5	7.1	2.5
	Water Usage (L)	190	60	190	60	190	60	190	60

FIGURE 9.24 Band saw results of the FEA and sustainability analyses.

		1		2		3		4	
	Design Type	Pin at Original Flywheel		Pin at Updated Original Flywheel		Pin at Different Flywheel		Pin at Updated Different Flywheel	
	Material	Steel, Low alloy	Steel, High Strength, Low Alloy	Steel, Low alloy	Steel, High Strength, Low Alloy	Steel, Low alloy	Steel, High Strength, Low Alloy	Steel, Low alloy	Steel, High Strength, Low Alloy
	Mass (kg)	0.189	0.192	0.189	0.192	0.189	0.192	0.189	0.192
Strength Analysis	Max. Von Mises Stress (MPa)	200.2	173.5	346.4	308.9	318.7	292.4	500.9	480.6
	Max. Displacement (mm)	0.04893	0.04068	0.08123	0.08341	0.07801	0.06871	0.09204	0.09054
	Min. Factor of Safety	1.25	1.59	0.79	0.89	0.78	0.94	0.5	0.57
Eco-material Sustainability Analysis	Energy Usage (MJ)	5.5	5.5	5.5	5.5	5.5	5.5	5.5	5.5
	CO_2 Footprint (kg)	0.37	0.38	0.37	0.38	0.37	0.38	0.37	0.38
	Water Usage (L)	9.4	9.1	9.4	9.1	9.4	9.1	9.4	9.1

FIGURE 9.25 Flywheel pin results of the FEA and sustainability analyses.

flywheel with the straight spoke, and flywheel with the straight spokes and longer hub, respectively. Stress and sustainability investigations were performed for each combination of the band saw with different flywheel designs and two different band materials. The results show that high-strength low-alloy steel provides better factor of safeties as well as better results in sustainability aspects when compared to austenitic stainless-steel ban material.

9.4.5 FEA of the Flywheel Pin

Flywheel pin was analyzed in a similar manner with the band saw. The only parameter of the flywheel pin that was varied is the material – Steel (High Strength Low Alloy) and Steel (Low Alloy). Images 1–4 in Figure 9.25 show responses in the form of von Mises stress distributions to the flywheel through the original design, with thicker spokes, with simplified straight spokes, and with straight spokes and longer hub, respectively. Regions of the flywheel pin that are colored in reddish colors are experiencing bending and shear stresses, whereas stresses are gradually reducing toward the free end of the pin. Yield strength was chosen instead of UTS to avoid plastic deformation. Most of the cases as shown in Figure 9.25 predict failure with factor of safeties below 1. The greatest value for minimum factor of safety turned out to be 1.59 for the selected pin working with the original flywheel, which prevents the use of improved models. Therefore, as a remedy, the flywheel pin diameter was increased to 30 mm which yields a minimum factor of safety of 1.71.

9.4.6 Results and Discussion

As a result of the comprehensive analyses of the critical parts of the band saw mechanism, an optimal combination of the flywheel, band saw, and flywheel pin was chosen as depicted in Figure 9.26, which is a different design set from the previous, however, emerged by using the design decisions set and tested in the prior trials.

Figures 9.27–9.29 illustrate the comparison of all results in the form of bar charts.

Figure 9.27 shows the bar chart that compares the flywheel designs in terms of minimum factor of safeties, CO_2 footprint, and energy usage. Bar no. 1 represents the originally designed flywheel, no. 2 representing the flywheel with thicker spokes, no. 3 representing the flywheel with simplified straight arms, and no. 4 being the flywheel design with straight spoke and longer hub. Bar no. 5 shows the results for the

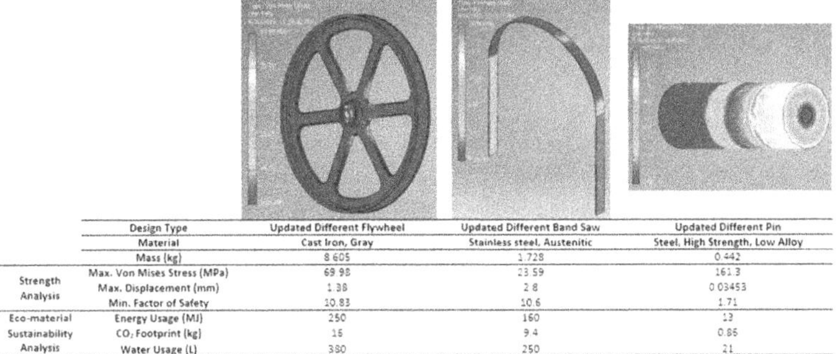

	Design Type	Updated Different Flywheel	Updated Different Band Saw	Updated Different Pin
	Material	Cast Iron, Gray	Stainless steel, Austenitic	Steel, High Strength, Low Alloy
	Mass (kg)	8.605	1.728	0.442
Strength Analysis	Max. Von Mises Stress (MPa)	69.98	23.59	161.3
	Max. Displacement (mm)	1.38	2.8	0.03453
	Min. Factor of Safety	10.83	10.6	1.71
Eco-material Sustainability Analysis	Energy Usage (MJ)	250	160	13
	CO_2 Footprint (kg)	15	9.4	0.86
	Water Usage (L)	380	250	21

FIGURE 9.26 Final optimal design decision.

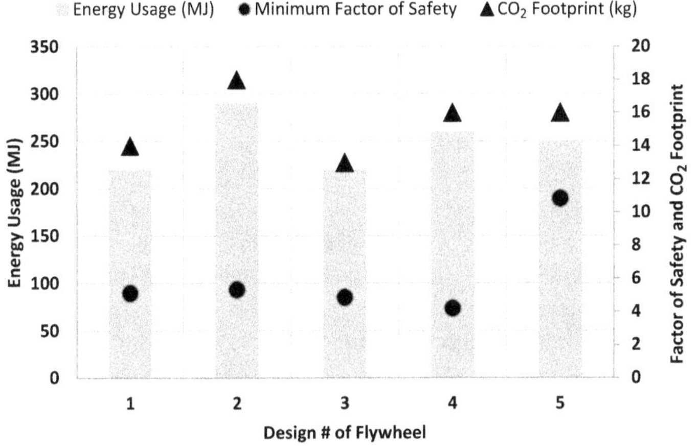

FIGURE 9.27 Comparison of flywheel designs in terms of minimum factor of safeties, CO_2 footprint, and energy usage – fifth being the optimal selection.

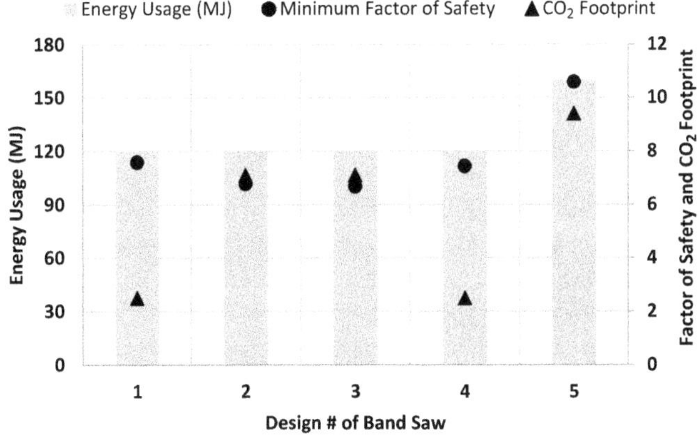

FIGURE 9.28 Comparison of band saw designs in terms of minimum factor of safeties, CO_2 footprint, and energy usage – fifth being the optimal selection.

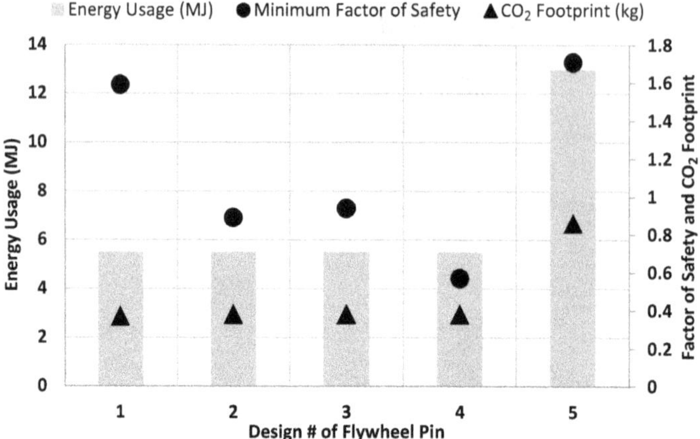

FIGURE 9.29 Comparison of flywheel pin designs in terms of minimum factor of safeties, CO_2 footprint, and energy usage – fifth being the optimal selection.

optimal flywheel design that produced the best combined results with the band saw and flywheel pin selections. In Figures 9.28 and 9.29, optimal band saw and flywheel pin selections are also shown with Bar No. 5. It can be concluded that the overall minimum factor of safeties of the newly designed reverse engineered band saw mechanism almost doubles the old results. Therefore, it can be claimed that machine tool can operate at higher speeds eliminating extra risk, or on the other hand, it can operate with the originally set speeds while the operation being much safer.

9.5 CONCLUSIONS

This chapter provides a framework for the deeper understanding of geometric and functional reverse engineering to assess and improve selected machine tools through computational modelling with an emphasis on sustainable and economical design. In the first case study, a bench-type drilling machine was reverse-engineered, as a result of which an improved model was proposed through the enhancement of the features of the most critical component of the drilling machine, namely, the spindle. Employing an evaluation matrix, seven different cases, consisting of the three spindle models and three different materials, were analyzed to determine the optimum solution for minimizing the stresses in the spindle and for reducing the carbon dioxide footprint and cost of the materials. The spindle model with 5 fins and Stainless Steel AISI 440C material provided the highest factor of safety, whereas the spindle model with 7 fins and Malleable Cast-Iron material indicated the most environmental-friendly design.

In the second case study, slightly different approaches were utilized in order to reverse-engineer and improve a band saw mechanism, where modifications on the critical machine elements of flywheel, band saw, and the flywheel pin were carried out. The responses in the form of maximum von Mises stress distributions, maximum displacements, and minimum factor of safeties are collected as well as results from sustainability analysis such as energy usage, carbon dioxide footprint, and water usage.

As a conclusion, an effective and feasible methodology is presented in this chapter that allows for improvement of available machinery in a fast, sustainable, and cost-effective manner. Within this methodology, reverse engineering plays a premier role coming forward as a promising and powerful solution to various similar manufacturing problems.

ACKNOWLEDGMENTS

Consideration and invitation for participation in the WRE 2020 and financial support from the Ghulam Ishaq Khan Institute of Engineering Sciences and Technology (GIKI), Pakistan, and financial and in-kind support from the Middle East Technical University – Northern Cyprus Campus (METU NCC), Turkey, are gratefully acknowledged.

REFERENCES

1. Ingle KA. *Reverse Engineering.* McGraw-Hill Professional Publishing, New York, 1994, pp. 1–35.
2. Murphy MD., Wood, KL., Otto, K., Bezdek, J., Jensen, D., Building better mousetrap builders: courses to incrementally and systematically teach design. *Paper presented at 1998 ASEE Annual Conference*, Seattle, Washington, DOI: 10.18260/1-2–6948, pp. 3.129.1–3.129.21, June 1998.
3. Tang D, Zhu R, Chen X, Zang T, Xu R. Functional reverse engineering for re-creation design. In *Proceedings of the 6th CIRP-Sponsored International Conference on Digital Enterprise Technology 2010* (pp. 185–195). Springer, Berlin, Heidelberg.
4. Hirtz J, Stone RB, McAdams DA, Szykman S, Wood KL. A functional basis for engineering design: Reconciling and evolving previous efforts. *Research in Engineering Design.* 2002 Mar 1;13(2):65–82.
5. Lin YP, Wang CT, Dai KR. Reverse engineering in CAD model reconstruction of customized artificial joint. *Medical Engineering & Physics.* 2005 Mar 1;27(2):189–93.
6. Yau, H.T., Reverse engineering of engine intake ports by digitization and surface approximation, *International Journal of Machine Tools and Manufacture*, 1997;37(6):855-871.
7. Sokovic M, Kopac J. RE (reverse engineering) as necessary phase by rapid product development. *Journal of Materials Processing Technology.* 2006;175(1–3): 398–403.
8. Pandilov Z, Shabani B, Shishkovski D, Vrtanoski G. Reverse engineering–An effective tool for design and development of mechanical parts. *Acta Technica Corviniensis-Bulletin of Engineering.* 2018 Apr 1;11(2):113–8.
9. Varady T, Martin RR, Cox J. Reverse engineering of geometric models—An introduction. *Computer-aided Design.* 1997 Apr 1;29(4):255–68.
10. Shimomura Y, Yoshioka M, Takeda H., et al. Representation of design object based on the functional evolution process model. *Transaction of the ASME - Journal of Mechanical Design.* 1998;120:221–29.
11. Fantoni GU, Taviani C, Santoro R. Design by functional synonyms and antonyms: A structured creative technique based on functional analysis. *Proceedings of the Institution of Mechanical Engineers, Part B: Journal of Engineering Manufacture.* 2007 Apr 1;221(4):673–83.
12. Cross N. *Engineering Design Methods: Strategies for Product Design.* John Wiley & Sons Inc. John Wiley & Sons, Chichester, 2000, pp 77–89.

13. Otto KN, Wood KL. Product evolution: A reverse engineering and redesign methodology. *Research in Engineering Design.* 1998 Dec 1;10(4):226–43.
14. Gietka P, Verma M, Wood WH. Functional modeling, reverse engineering, and design reuse. In *International Design Engineering Technical Conferences and Computers and Information in Engineering Conference.* 2002 Jan 1 (Vol. 3624, pp. 207–18).
15. Kirschman CF, Fadel GM. Classifying functions for mechanical design. *Transaction of ASME- Journal of Mechanical Design.* 1998;120(3):475–82.
16. Mahboubkhah M., Aliakbari M., Burvill, C., An investigation on measurement accuracy of digitizing methods in turbine blade reverse engineering, *Proceedings of the Institution of Mechanical Engineers, Part B: Journal of Engineering Manufacture.* 2018;232(9):1653–1671.
17. Yau HT, Menq CH. Automated CMM path planning for dimensional inspection of dies and molds having complex surfaces. *International Journal of Machine Tools and Manufacture.* 1995 Jun 1;35(6):861–76.
18. Jones SL. Laser digitizes 3-D model data for CAD/CAM. *Metalworking News.* 1989 Apr 17;12.
19. Farnum GT. Measuring with lasers. *Manufacturing Engineering.* 1986 Feb 1;96(2):47–51.
20. Lee KH, Park H, Son S. A framework for laser scan planning of freeform surfaces. *The International Journal of Advanced Manufacturing Technology.* 2001 Jan 1;17(3):171–80.
21. JET Tools, 2019. VBS-18MW, 18″ Metal/Wood Vertical Band saw. [Online] Available at: https://www.jettools.com/us/en/p/vbs-18mw-18-metal-wood-vertical-bandsaw/414418 [Accessed 7 Apr. 2019].

10 A Coarse-to-Fine Classification Method for Aviation High-Similarity Sheet Metal Parts Based on Machine Vision

Lv Zhengyang
Nanjing University of Aeronautics and Astronautics

Li Xiaojun
AVIC Chengdu Aircraft Industrial (Group) Co. Ltd.

Lin Yutao and Zhang Liyan
Nanjing University of Aeronautics and Astronautics

CONTENTS

10.1 INTRODUCTION

Sheet metal parts (hereafter referred to as SMP) have extensive applications in aviation manufacturing. According to a rough estimate, there are more than several tens of thousands of types of SMPs in an aircraft body, which account for about 20%–40% of all the structural parts. Aviation SMPs have numerous varieties and are usually formed in small batches. After a certain number of SMPs are formed, they need to be painted in batch. In the painting process, different kinds of SMPs are mixed. In Figure 10.1, 20 types of SMPs are demonstrated, each of them with a very similar symmetry counterpart. Reidentification of these parts after painting is a tough task in practice. At present, the classification of high similarity SMPs can only be carried out by manual comparison in practice, which is not only laborious and time-consuming but also prone to classification errors.

Thanks to the continuous efforts of scientific and industrial communities, machine vision has achieved considerable progress [1]. As an important part of machine vision, image-based classification has been extensively investigated. Excellent image classification algorithms, such as those based on the principles of word packet [2], template matching [3], and sparse representation [4], have been well applied in various industry fields [5]. Among others, Joshi and Surgenor [6] combined shape descriptors [7] such as the average intensity and circularity of images in their algorithm to represent the part images. Cusano and Napoletano [8] utilized matched templates [3] and local descriptors to localize and recognized 8 panels and 20 parts from 248 images taken from a real aircraft. The scale-invariant feature transform (SIFT) algorithm [9], which has been widely applied in various computer vision tasks, assigns a 128-dimensional vector as the local descriptor for each extracted key point in an image. Li et al. [10] proposed an automatic method for detecting the brake-shoe-key losing of freight trains. They fused 11 features like contour smoothness and

FIGURE 10.1 Different kinds of SMPs.

contour concave-convex [11] to represent the objects to be classified. The support vector machine (SVM) [12] and artificial neural network [13] are the most popularly used classifiers in various applications.

The emergence of convolutional neural network (CNN) in recent years has brought radical changes in image-based classification [14]. A series of long-standing problems has been successfully solved by using CNN [15–16]. Among others, Yin et al. [17] built a CNN named as Part R-CNN, which is based on Faster R-CNN [18], to recognize the assembly parts. The network takes the images of sample electromechanical products as input and output the corresponding part types. They reported an average recognition precision of 94% and a high recognition speed in their experiments.

Even though great progresses have been made in object classification, it still needs to verify whether the state-of-the-art classification methods are effective for SMPs. Since there are huge kinds of aircraft SMPs and some of them involve very high similarity as shown in Figure 10.1, the classification of SMPs is a hard task. In addition, the characteristics of SMPs, such as lack of texture and no reliable locating datum, make the classification even harder. Still worse, because of the huge workload it might involve, to collect many sample images for each type of SMPs is impossible in practice, whereas large sample data set is indispensable to statistic-learning-based classification methods. To the best of our knowledge, no reports have been published on classification of aircraft SMPs.

In this chapter, we first conduct a preliminary study to examine the applicability of several popular object classification methods on the SMPs. The performance of the trialed methods is reported, which demonstrates that the features such as scale-invariant feature transform (SIFT) and histogram of oriented gradient (HOG) are not discriminative enough for the classification of SMPs. Then, a machine-vision-based coarse-to-fine classification scheme is put forward. In this scheme, two industrial cameras are fixed on a working platform to take images from the top and the side views of each SMP. The sample images of all type parts, together with the ten-dimensional (10D) feature vectors extracted from each of the top view images, are stored in a database. Other than the local features popularly used, the 10D translation and rotation invariant feature vector that we exploit represents global shape characteristic of the SMPs. It proves to be adequate for the coarse-grained classification, which retrieve the two most similar candidates from the database. The coarse-grained classification result was utilized to guide the operator to relocate the part on the working platform to largely the same position and orientation as that of the candidates in the database. In this way, we have effectively realized repetitive positioning of the SMP, which is of great importance for the fine-grained classification. Then the contour curve similarities between the top and the side view images of the candidates and that of the part in trial were computed and were used to achieve fine-grained classification. The classification process is shown in Figure 10.2.

The rest of the chapter is structured as follows. In Section 10.2, three popular classification methods are implemented and their performances on classification of the SMPs are reported. The proposed coarse-to-fine classification method is depicted in Section 10.3. Finally, we summarize our work and highlight some promising directions for future work in Section 10.4.

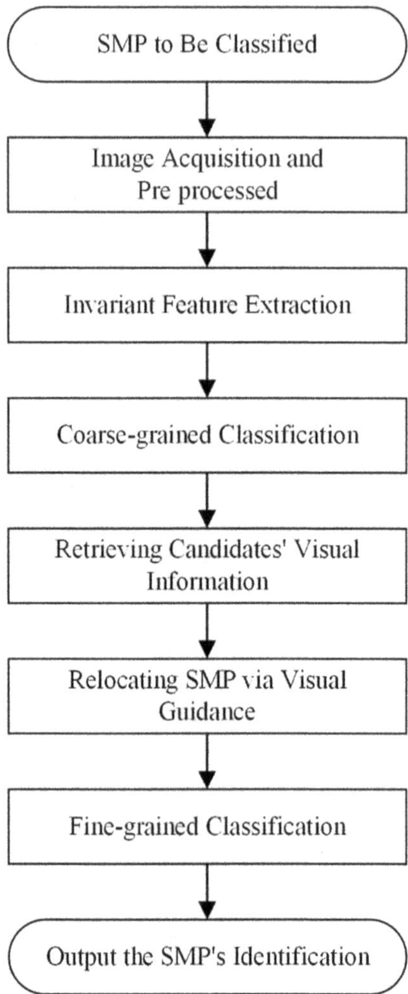

FIGURE 10.2 Classification process.

10.2 PRELIMINARY STUDY

Considering that SVM and CNN have exhibited excellent performance in various object classification tasks, we first examine their applicability for the classification of SMPs in this section.

10.2.1 Brief Introduction to the Trialed Methods

Image-based classification methods could be divided into two categories according to the ways of feature extraction. One category is based on hand-crafted features, such as SIFT [9], HOG [19], and Shape Descriptors [7]. The other category is based on deep learning and CNN, which automatically finds features from the raw data [14].

In this section, three methods from both categories, namely SIFT + SVM, HOG + SVM, and the state-of-the-art CNN VGG-16 [20], are realized and tested for the classification of SMPs.

SIFT: The SIFT descriptor is scale and rotation invariant and light insensitive, which shows strong abilities in object recognition. It has been widely used in local feature detection and image matching. The first two steps of traditional SIFT method are detecting scale-space extrema and localizing the key points. To speed up the algorithm, 1000 patches from each image are randomly sampled as the locations and extract SIFT descriptors from the patches as in Ref. [21]. For all images, 1000 element codebooks are built with online k-means and mutual-information-based histogram encoding is used with an SVM classifier.

HOG: HOG is a global feature descriptor firstly applied in the field of human detection and popularized in general objects detection. There are several advantages of the HOG descriptor. It captures edges or gradient structures that are very characteristic of local shape, and it does so in a local representation with an easily controllable degree of invariance to local geometric and photometric transformations [19]. Each pixel's gradient is computed and accumulated into orientation bins over spatial regions which called as cells. Cells group together into larger spatial blocks and each block is normalized separately. Compared with local descriptor which puts effort into determining the key points' location, HOG descriptor is the vector of all components of the normalized cell responses from all the blocks in the image. In the experiments, a 3780-dimensional HOG descriptor was used as in Ref. [19] to feed the SVM for classification.

SVM: Because of its superiorities in classification tasks, SVM has been adequately developed and applied in industrial field for a long time. Essentially, this method finds a hyperplane in space, which could distinctively divide all data samples and ensure that the distance from all data to this plane is the minimum [12]. The method transforms the classification into the convex quadratic programming task. The distance between each data point and the hyperplane could be calculated, which is computed by the general expression at first. Then, support vectors are defined as data points which are on or closest to the hyperplane. The core of the algorithm is identifying the hyperplane that has the maximum distance between the nearest data points of either class. This distance is called as margin. By solving the given constraint conditions, the maximum margin could be found and corresponding hyperplane could be obtained. The main disadvantage of SVM is that each model of SVM could only classify certain number of part categories. In other words, it needs to refresh the model whenever adding a new kind of part's data.

CNN: CNN showed its strong abilities in the field of image classification [14]. A CNN is usually composed of convolution layers, pooling layers, and a full connection layer. There are two stages in CNN-based classification. First, feature extraction is completed in the convolution layers and the pooling layers. The convolution layers oversee representing the characteristics of the original image. However, the pooling layers are designed to reduce the dimensionalities and training parameters. Second, the full connection layer is utilized to complete the classification task. In the full connection layer, the feature matrix is divided into a series of vectors. After a series of processing, such as multiplying coefficient matrix, adding bias, and using activation function, a classifier with fixed number of outputs is built. The state-of-the-art CNN, namely VGG-16, was used in this work.

10.2.2 Preliminary Experiment Results

There were 400 sample images collected by capturing 20 pictures for each of the 20 types of SMPs as shown in Figure 10.1. One hundred images were randomly selected as the test set. Cross-validation was utilized in all experiments' training phase. Among the other 300 images, 220 images were taken as the training set and 80 images were involved in the validation set in the experiments of SIFT + SVM and HOG + SVM. To enhance the performance of the CNN algorithm, data augment technology [22] was used. More specifically, 2400 training images were synthesized according to the 300 images in the original training set and validation set. The sizes of the test set, the training set, and the validation set were kept unchanged in the experiments. Top 1 Score and Top 2 Score were selected to evaluate their performance. The Top 1 Score checks if the one with the highest probability is the right target. The Top 2 Score checks if the true target is within ones with the top 2 highest probabilities.

Referring to Table 10.1, the top 1 scores of the tested methods on SMP classification were all below 70%, which is far away from practical requirement. It is obvious that the hand-crafted descriptors are not fully competent for representing the image characters of the SMPs. They are inevitably to be overfitting no matter how skillfully the parameters are refined. VGG-16 has strong ability in analysis abstract and high-dimensional features for general object detection; however, it still cannot achieve satisfactory result on SMP classification, even though the sample data were augmented to a larger set. Besides, it is hard to update the classification model once the training process in CNN or SVM is finished. In other words, when additional new types of SMP need to be classified, the whole time-consuming training process should run once again. The total time cost of descriptor extraction in the first two experiments and model training in VGG-16 were all more than 1 hour. From Figure 10.1 we can observe that each of the SMPs in practice (also in our experiments) has a very similar counterpart. Each similar pair is relatively easy to be distinguished from the others. This observation well explains the experimental result that the Top 2 Score of HOG + SVM algorithm (90.0%) is much higher than the Top 1 Score (48.0%). More specifically, the 90.0% Top 2 Score indicates that HOG + SVM algorithm has successfully distinguished most similar pairs from the others, while Top 1 Score of less than 50% indicates it totally fails in differentiating the two SMPs in similar pairs.

TABLE 10.1
The Results of Tested Classification Methods

Methods	SIFT + SVM	HOG + SVM	CNN
Top 1 score (%)	66.0	48.0	65.0
Top 2 score (%)	71.0	90.0	82.0

The Top 2 Score of SIFT+SVM (71.0%) is much less than that of HOG+SVM (90.0%), which demonstrates that the global feature HOG outperforms the local feature SIFT in separating the similar pairs from each other.

10.3 COARSE-TO-FINE CLASSIFICATION METHOD

Considering the pairwise character of the SMPs, a coarse-to-fine vision classification method was proposed to balance the accuracy and efficiency requirements. In the sample data collection stage, two images of each type of SMPs in consideration are taken, respectively, from the top and the side views by two industrial cameras fixed on a working platform. The sample images of all type SMPs, together with the feature vectors extracted from each of the top view images, were stored in a database. In the part classification stage, the part in trial was placed on the working platform and two images were taken just as in the data collection stage. The classification method splits the original N-to-1 task into two tasks, namely, an N-to-2 task and a 2-to-1 binary classification task. In the N-to-2 task, i.e., the coarse-grained classification, the top 2 candidates from the database were found. While in the 2-to-1 task, i.e., the fine-grained classification, the correct one was determined from the highly similar two candidates.

10.3.1 Image Acquisition and Preprocessing

An image collection platform as shown in Figure 10.3 was built. Two CMOS cameras of AVT Mako g-158b PoE, each with a Schneider Kreuznach 1.4-8 and a Kowa LM12NCL 1.4-12 industrial lens, respectively, were fixed with the platform. The optical axis of one camera, which will be called as the main camera hereafter, was perpendicular to the working platform. The other camera, called as assistant camera

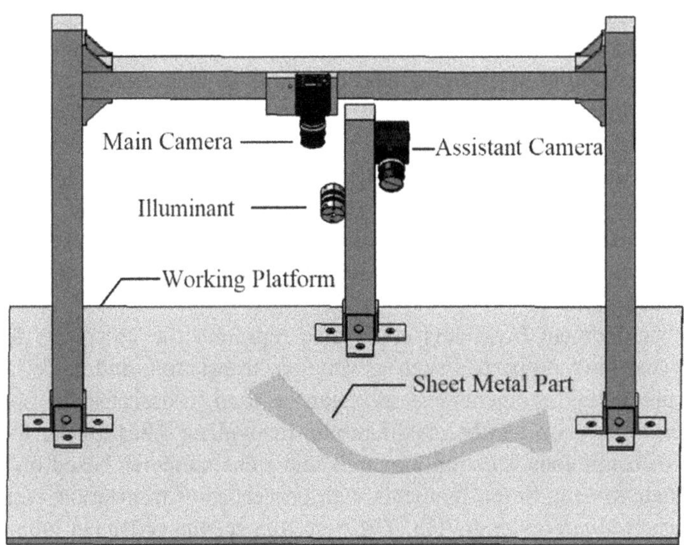

FIGURE 10.3 Schematic diagram of our image collection hardware.

hereafter, takes picture from a side perspective. Instead of exploiting large numbers of sample images to train a classification model, each type of the SMPs was represented by using only two sample images captured by the main and assistant cameras at once. The two images are also referred to as the top-view image and the side-view image, denoted as I^T and I^S, respectively. For the two sample images of each type of SMPs, Gaussian smooth [23] was first performed to denoise the images, then the images were binarized with a thresholding process. The top and side contours of the SMP, denoted as C_T and C_S, were extracted from the binary images. A database was established with the abilities of storing, indexing, and retrieving the original images I^T and I^S, the 10D feature vector F (see the next section for detail), top and side contours C_T and C_S, and the corresponding identification ID of every SMP. When an SMP was to be classified, it was placed on the working platform and two images were taken just as in the sample image collection stage. In the following sections, the coarse-to-fine method is described in detail.

10.3.2 Coarse-Grained Classification

Global shape descriptors were used to represent the coarse feature of the SMPs. The coarse-grained classification was conducted by comparing the relative Euclidean distance between each feature vector in the database and the feature vector of the part to be classified.

Shape Descriptors: Shape Descriptors generally look for effective and perceptually important shape features based on either shape boundary information or boundary plus interior content [7]. In other words, Shape Descriptors could be generally divided into two classes: contour-based methods and region-based methods. The global shape descriptors were selected, namely, Shape Invariants and Geometric Moment Invariants, in the top-view image I^T to form a 10D feature vector for each SMP.

Shape Invariants can describe the inherent features of graphics. They have advantages such as intuition, rotation and translation invariance, low calculation cost, and strong depicting ability toward graphics. There are six Shape Invariants used in this method, including the shape's perimeter Pe and area A. By fitting an ellipse to the shape [24], the lengths L_a and S_a of the long axis and the short axis, respectively, were obtained. The radius of minimum circumscribed circle M_c and minimum bounding box area M_{bb} were also computed by using the efficient method in Ref. [25]. Since the image plane of the main camera and the work platform plane are parallel, these six factors $\{Pe, A, L_a, S_a, M_c, M_{bb}\}$ are invariant to the translation and rotation changes of the part on the platform.

Geometric Moment Invariants could also represent the geometric features of the image and have properties such as rotation, translation, and scale invariance. In image processing, moment invariants can be used to describe the objects, and from which the objects can be classified and recognized. Hu [26] put forward the definition of continuous function moments and rules moments based on rectangular coordinate system. Seven moments with properties of translation, rotation, and scale invariance were given by Hu. The first- and second-order Hu moments were used in this work, denoted as Hu_1 and Hu_2. Similar with the Hu moments, Zernike moments also have the characteristics of rotation translation and scale invariance.

One advantage of Zernike moments is that the expressions for any order moments can be constructed [27]. However, high-order Zernike moments are sensitive to noise. As a result, the zero- and second-order Zernike moments, marked as Z_{00} and Z_{20}, were used in this method.

The first- and second-order Hu moments can be calculated from the central normalized moments η_{pq} as Equation (10.1) to Equation (10.5), where $p,q = 0,1,2$.

$$\mu_{pq} = \sum_{y=1}^{N}\sum_{x=1}^{M}\left(x-\bar{x}\right)^{p}\left(y-\bar{y}\right)^{q} f(x,y) \tag{10.1}$$

$$\eta_{pq} = \mu_{pq} / \mu_{00}^{\rho} \tag{10.2}$$

$$\rho = \left[(p+q)/2\right]+1 \tag{10.3}$$

$$\text{Hu}_1 = \eta_{20} + \eta_{02} \tag{10.4}$$

$$\text{Hu}_2 = \left(\eta_{20} - \eta_{02}\right)^2 + 4\eta_{11}^2 \tag{10.5}$$

Equations (10.6)–(10.8) express the computation of zero- and first-order Zernike moments from spatial moments m_{pq}.

$$m_{pq} = \sum_{y=1}^{N}\sum_{x=1}^{M}\left(f(x,y)x^{p}y^{q}\right) \tag{10.6}$$

$$Z_{00} = \frac{1}{\pi} m_{00} \tag{10.7}$$

$$Z_{20} = \frac{6}{\pi}\left(m_{20} + m_{02}\right) - \frac{3}{\pi} m_{00} \tag{10.8}$$

Here, $f(x,y)$ denotes the image intensity at pixel (x,y). Having obtained the above ten parameters $\{Pe, A, L_a, S_a, M_c, M_{bb}, \text{Hu}_1, \text{Hu}_2, Z_{00}, Z_{20}\}$, a 10D feature vector F_i was constructed for the coarse-grained classification, where $i = 1,2,\ldots,N$, and N represents the total types of SMPs in the database. For the convenience of later expression, it is denoted as $F_i = \left(x_1, x_2, \ldots, x_{10}\right)^T$, where $x_1, x_2, \ldots, x_{10}$ stand for different feature components. As mentioned in the last section, the 10D feature vectors of the sample SMPs were stored in the database.

Nearest Neighbor Searching: Let P_x denotes the SMP to be classified and F_x denotes the feature vector of P_x. Then the relative Euclidean distance D_i between F_x and all the feature vectors F_i in the database was calculated one by one. Equation (10.9) expresses the calculation forum of D_i.

$$D_i = \left\|F_i - F_x\right\| = \sqrt{\sum_{j=1}^{n}\left(\frac{x_j - y_j}{y_j}\right)^2} \tag{10.9}$$

where x_j and y_j denote the jth component of F_i and F_x, respectively, and n represents the total number of components in the feature vector. In our case, $n = 10$. The coarse-grained similarity metric is defined as in Equation (10.10).

$$Sc_i = 1 - \frac{D_i}{\sum D_i} \tag{10.10}$$

Ideally, the relative Euclidean distance D_i should be close to 0 and the similarity metric Sc_i should be approximately equal to 1 if the ith SMP in the database is the true target. However, because of the high similarity exhibited in the pairwise SMPs, it is hard to determine the unique true identification for the part to be classified via comparing the 10D feature vectors. Therefore, two SMPs in the database, whose similarity metric Sc are the top 2 largest ones, were selected as the candidates. The two candidate targets, denoted as T_0 and T_1 respectively, would be further classified in the following steps.

10.3.3 Visual-Guided Part Relocation

The top and side contours C_T and C_S of the two candidate targets T_0 and T_1 in the database were taken as input data in the fine-grained classification. The contour C_S in the side-view image is sensitive to the relative location between the assistant camera and the SMP. Unfortunately, the lack of reliable locating datum for SMPs makes it hard to repeat the location of the same part. In order to utilize C_S to further determine whether T_0 or T_1 is the unique right answer, a visual guidance for the operator was provided to relocate the part on the working platform to largely the same position and orientation was highlighted as that of the candidates in the database. More specifically, the part boundary increases the transparency of the interior of T_0 and T_1. An example image processed in such a way is shown in Figure 10.4. The visually processed side-view images of T_0 and T_1 were then, respectively, overlapped with the

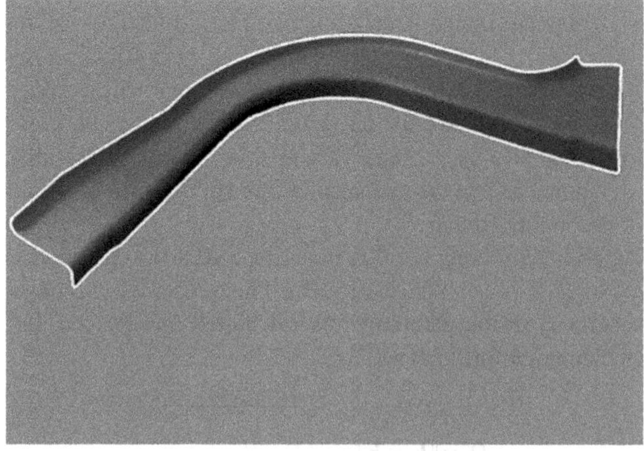

FIGURE 10.4 The processed side-view image of a candidate in the database.

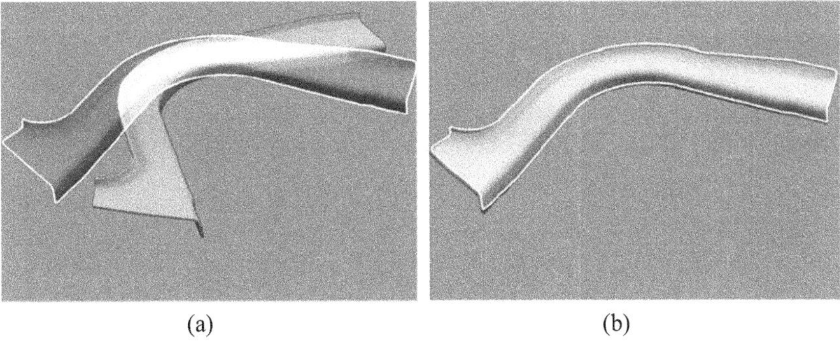

(a) (b)

FIGURE 10.5 One synthesized image by overlapping the real-time video image and the visual guidance image: (a) relative position before relocation and (b) relative position after relocation.

real-time captured side-view image of P_x, as shown in Figure 10.5a. The two synthetized images were displayed on the interface of the classification software. With the visual guidance from the synthetized images, the operator can easily adjust the position and orientation of P_x on the platform until the boundary of P_x largely coincides with that of T_0 or T_1, as shown in Figure 10.5b. In this position, P_x is recaptured as two images, denoted as I_x^T and I_x^S, respectively, by the main and the assistant cameras simultaneously for the following fine-grained classification.

10.3.4 Fine-Grained Classification

In the fine-grained classification, the boundaries of P_x were first extracted from the images I_x^T and I_x^S by using the method in Ref. [28]. Here, the extracted boundaries are denoted as C_x^T and C_x^S. The Iterative Closest Point (ICP) algorithm [29] was then used to, respectively, align the boundaries C_x^T and C_x^S with the corresponding boundaries of the candidates T_0 and T_1. Considering the actual situation in this context, a 2D version ICP algorithm was implemented, which is much more computationally efficient than the commonly used 3D version.

After the corresponding boundaries were best aligned via ICP algorithm, the total Root Mean Square (RMS) distance was calculated as defined in Equation (10.11) to measure the difference between P_x and the two candidates T_0 and T_1.

$$\text{RMS}_l = d_{\text{RMS}}\left(C_x^T, C_l^T\right) + d_{\text{RMS}}\left(C_x^S, C_l^S\right), l = 0,1. \quad (10.11)$$

Here $d_{\text{RMS}}(\cdot,\cdot)$ represents the RMS distance between two boundary contours, C_l^T and C_l^S denote, respectively, the boundary contours in the top or side view image of the candidate T_l. To normalize the fine-grained similarity metric to 0~1, Sf_l is defined as in Equation (10.12).

$$Sf_l = 1 - \frac{\text{RMS}_l}{\sum_l \text{RMS}_l} \quad (10.12)$$

To verify the performance of visual guidance, there is one more constraint before predicting the target object. The fine-grained classification only works when at least one of the RMS_l is less than 20, which is a general practice. If not, P_x needs to be relocated until position and orientation are as same as possible as that of the candidates in the database. At last, the candidate whose Sf_l score is larger than the other one is taken as the final Top 1 classification result T_f.

10.3.5 Coarse-to-Fine Classification Results

In the sample data collection stage, the main and the assistant cameras, respectively, take an image for each of the 20 sample SMPs as shown in Figure 10.1. The 40 images and extracted boundaries and feature vectors were recorded in the database. Implementations were made in Visual Studio 2017 using the library of OPENCV. A computer with Intel Core i5 7400 processor and Windows 10 operating system were used in the experiments. The coarse-grained and the fine-grained algorithms were written in C++ programing language and the pseudocodes are shown in Algorithms 1 and 2.

Algorithm 1: Coarse-Grained Classification

Input: The top-view image I_x^T of SMP to be classified.

Extract F_x from I_x^T

Obtain the number of samples SMPs in the database, denoted as S.

For $i = 1, 2,\ldots, S$

Calculate the Euclidean distance $D_i = \|F_i - F_x\| = \sqrt{\sum_{j=1}^{n}\left(\frac{x_j - y_j}{y_j}\right)^2}$

For $i = 1,2,\ldots, S$

Calculate the coarse-grained similarity $Sc_i = 1 - \frac{D_i}{\sum D_i}$

Find out the 2 candidate parts in the database T_l, $l = 0,1$, which has the two largest Sc values.

Output: Two candidate SMPs T_l, $l = 0,1$

Algorithm 2: Fine-Grained Classification

Input: Candidates SMP T_l, the top-view image I_x^T and side-view image I_x^S from the SMP to be classified

Extract the boundary contours C_l^T and C_l^S in the top or side view image of the T_l from database.

Extract the boundary contours C_x^T and C_x^S in the top-view image I_x^T and side-view image I_x^S of the SMP to be classified

For $l = 1,2$

Calculate the RMS distance between two boundary contours $\text{RMS}_l = d_{\text{RMS}}(C_x^T, C_l^T) + d_{\text{RMS}}(C_x^S, C_l^S)$

For l = 1,2

Calculate the fine-grained similarity $Sf_l = 1 - \frac{\text{RMS}_l}{\sum_l \text{RMS}_l}$

Find out the maximum Sf and corresponding sample part in the database as the final result T_f

In classification stage, the main camera was first used to take one image for the unknown SMP on the platform and test the performance of the coarse-grained classification. The coarse-grained classification experiments were performed 100 times in total. The SMPs being classified in the experiments include all different kinds of SMPs in the database. The Top 1 Score and Top 2 Score of the coarse-grained classification were 53% and 100% separately. It took only 1.05 seconds on average to get a coarse-grained classification result. It is worth of notice that in the coarse classification, the unknown part can be freely put on the platform if only it is within the field of view of the main camera. No relocation is necessary in this stage.

In Table 10.2, some examples of the feature vectors are presented and the relative Euclidean distances calculated in the experiments. The part in Column 2 is the unknown to be classified. The third and the fourth columns illustrate, respectively, the true target and its high similar counterpart in the database. It can be seen in the 13th row of Table 10.2, the distances between the unknown and the SMPs in the third and the fourth columns were 0.002 and 0.059, respectively, which are obviously smaller than the other distances. Therefore, the coarse-grained classification method can reliably find out the two most similar candidates of the unknown. On the other hand, the coarse-grained classification cannot confidently determine which candidate is the unique correct answer, since the normalized Sc scores of the two candidates (99.99% and 99.77%) are too close to be distinguished.

To verify the performance of the fine-grained classification, each of the 100 coarse-grained experiments following the methods in the last two subsections to determine the final unique answer. The example fine-grained similarity metrics of the SMP pairs are listed in Table 10.3. We can see that most SMPs' Sf scores of the true targets are higher than 92.0%, while the Sf scores of the false candidate targets are lower than 8.0%. However, the SMP shown in the last column is hard to be classified by our method. There is slight gap among these two parts in practice, to be exactly, only 2 mm difference in thickness and it is nearly impossible to classify by image-based methods. In Figure 10.6, the effects of contour alignment with ICP algorithm are illustrated. The two cases in Figure 10.6a illustrate the contour alignment between parts of the same kind, from which we can see that each pair of contours almost coincide together after alignment. The two cases in Figure 10.6b demonstrate the contour alignment result of two different parts. The difference between the two contours after alignment is still much larger. In the 100 fine-grained classification experiments, 96 correct predictions were obtained. In other words, the coarse-to-fine strategy for SMP classification finally achieves an accuracy of 96% in this experiment. The average running time of all the experiments was counted,

TABLE 10.2
Feature Data and Relative Distances between the Unknown and Eight Categories of SMPs in the Database

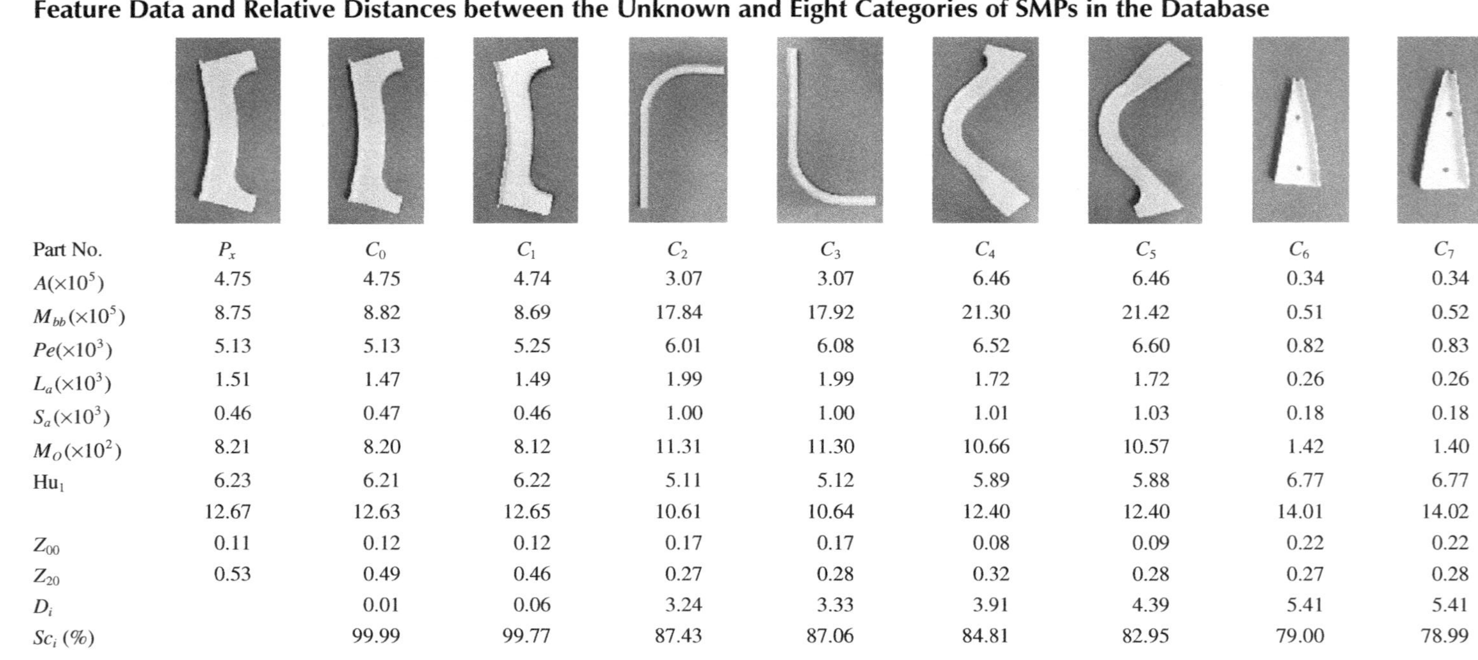

Part No.	P_x	C_0	C_1	C_2	C_3	C_4	C_5	C_6	C_7
$A(\times10^5)$	4.75	4.75	4.74	3.07	3.07	6.46	6.46	0.34	0.34
$M_{bb}(\times10^5)$	8.75	8.82	8.69	17.84	17.92	21.30	21.42	0.51	0.52
$Pe(\times10^3)$	5.13	5.13	5.25	6.01	6.08	6.52	6.60	0.82	0.83
$L_a(\times10^3)$	1.51	1.47	1.49	1.99	1.99	1.72	1.72	0.26	0.26
$S_a(\times10^3)$	0.46	0.47	0.46	1.00	1.00	1.01	1.03	0.18	0.18
$M_O(\times10^2)$	8.21	8.20	8.12	11.31	11.30	10.66	10.57	1.42	1.40
Hu_1	6.23	6.21	6.22	5.11	5.12	5.89	5.88	6.77	6.77
	12.67	12.63	12.65	10.61	10.64	12.40	12.40	14.01	14.02
Z_{00}	0.11	0.12	0.12	0.17	0.17	0.08	0.09	0.22	0.22
Z_{20}	0.53	0.49	0.46	0.27	0.28	0.32	0.28	0.27	0.28
D_i		0.01	0.06	3.24	3.33	3.91	4.39	5.41	5.41
Sc_i (%)		99.99	99.77	87.43	87.06	84.81	82.95	79.00	78.99

TABLE 10.3
Computed Similarity Metrics of Ten Pair SMPs

Figures			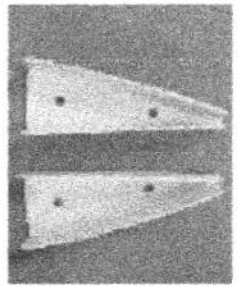				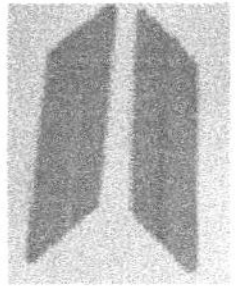			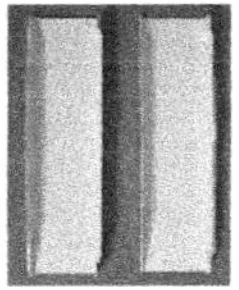	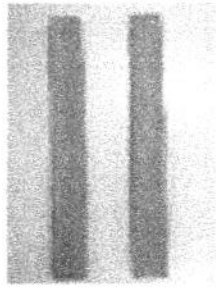
The True Target	RMS	0.04	0.73	13.00	14.65	10.00	3.11	0.32	2.96	5.22	16.50
	Sf (%)	99.89	98.23	92.63	98.04	95.21	98.44	99.86	99.09	93.68	46.68
The Similar but False Target	RMS	39.02	40.49	163.43	734.46	198.62	196.20	227.45	320.02	77.36	14.44
	Sf (%)	0.11	1.77	7.37	1.96	4.79	1.56	0.14	0.91	6.32	53.32

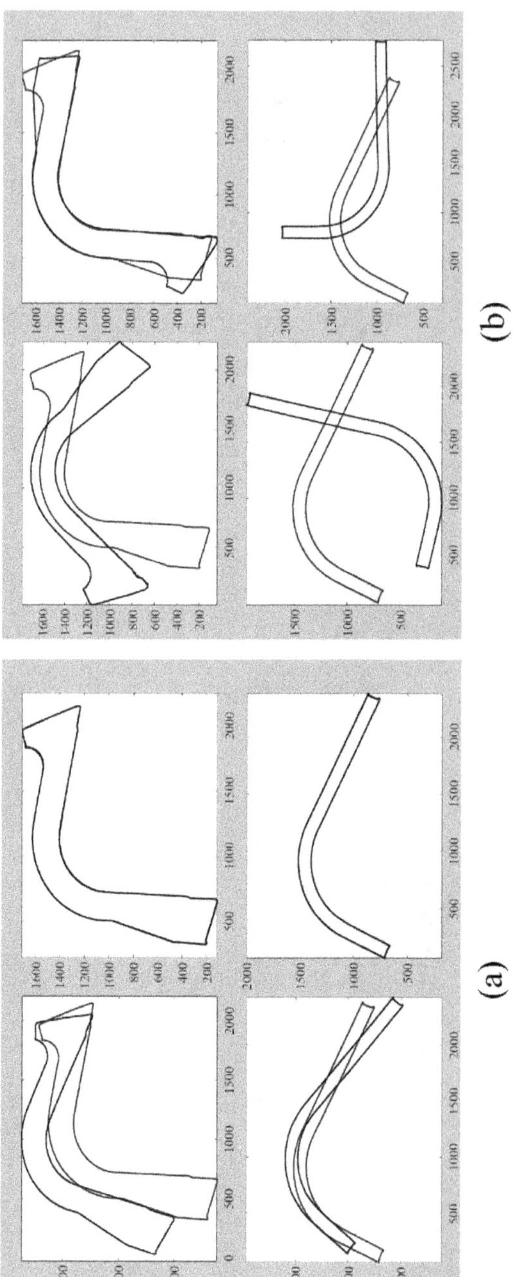

FIGURE 10.6 Illustration of contour alignment with ICP algorithm. (a) Alignment with the true target. (b) Alignment with the similar but false target. In both (a) and (b), the first column illustrates the prealignment state, and the second column illustrates the aligned state.

TABLE 10.4
The Experimental Performance of Our Method

Method	Coarse Only	Coarse + Fine
Time cost (s)	1.05	2.35
Top 1 score (%)	53.0	96.0
Top 2 score (%)	100.0	100.0

which does not include the time spent for visual-guidance relocation. The relocation procedure, depending on specific SMP and the operators' skill, usually takes less than 1 minute. Even coarse-to-fine classification method takes a little longer than the coarse method, it still meets up the real-time requirement in practices. Experimental performance of this method is concluded in Table 10.4.

10.4 CONCLUSION

A coarse-to-fine scheme to classify aviation SMPs based on machine-vision method has been presented in this chapter. An image collection platform with two industrial cameras to take images from different perspectives was built. The images extracted contours and 10D feature vectors are stored in a database. A 10D feature vector was utilized for the coarse classification, while in the fine classification, the top and the side contours were compared of the unknown SMP with that of the two candidates suggested by the coarse classification algorithm. A visual guidance strategy was put forward for relocating the unknown SMP to proper position and orientation before the fine-grained classification. It proves that this strategy is very effective. Hundreds of experiments on 20 different kinds of SMPs are conducted, and a classification accuracy of 96.0% is achieved with our coarse-to-fine method. Averagely, it takes only 2.35 seconds to recognize an SMP from the database in our experiments.

Instead of utilizing a huge number of sample data to train a classification model, the proposed method needs only two images for each type of SMPs. Compared to the popular classification methods based on statistic-learning, our method is far more efficient in both space and computation cost. More importantly, the proposed method exhibits much higher classification accuracy in the experiments.

The proposed method is of high scalability. Whenever a new additional type of SMP needs to be classified, it only needs to take two images of a sample part on the work platform and add a record in the database. Sure, how severely the efficiency and classification accuracy could be influenced with the increase of the record in the database is still up to comprehensive experiments.

There are still some limitations existing in our scheme. The progress of the scheme is step-by-step, which means the misclassification in the coarse phase would directly influence the results. And the quantity of sample parts in our database is not sufficient in fully verifying the performance of our scheme in practice. In the future, more types of SMPs will be recorded in the database, and more experiments will be conducted to validate the accuracy and efficiency of the proposed coarse-to-fine method. More features with geometric invariance could be added in the first phase, which will benefit the capacity of the coarse classification.

REFERENCES

1. Shapiro LG and Stockman GC. *Computer Vision [M]*. Prentice Hall Inc., New York, 2003.
2. Csurka G, Dance C, Fan L, Willamowski J, Bray C. Visual categorization with bags of keypoints. In *Workshop on Statistical Learning in Computer Vision*, ECCV 2004 May 15 (Vol. 1, No. 1–22, pp. 1–2).
3. Korman S, Reichman D, Tsur G, Avidan S. Fast-match: Fast affine template matching. In *Proceedings of the IEEE Conference on Computer Vision and Pattern Recognition*, IEEE 2013 (pp. 2331–8), New York, United States..
4. Wright J, Ma Y, Mairal J, Sapiro G, Huang TS, Yan S. Sparse representation for computer vision and pattern recognition. *Proceedings of the IEEE*. 2010 Apr 29;98(6):1031–44.
5. Cuevas E, Sossa H. A comparison of nature inspired algorithms for multi-threshold image segmentation. *Expert Systems with Applications*. 2013 Mar 1;40(4):1213–9.
6. Joshi KD, Surgenor BW. Small parts classification with flexible machine vision and a hybrid classifier. In *2018 25th International Conference on Mechatronics and Machine Vision in Practice (M2VIP)*, 2018 Nov 20 (pp. 1–6). IEEE.
7. Zhang D, Lu G. Review of shape representation and description techniques. *Pattern Recognition*. 2004 Jan 1;37(1):1–9.
8. Cusano C, Napoletano P. Visual recognition of aircraft mechanical parts for smart maintenance. *Computers in Industry*. 2017 Apr 1;86:26–33.
9. Lowe DG. Distinctive image features from scale-invariant keypoints. *International Journal of Computer Vision*. 2004 Nov 1;60(2):91–110.
10. Li N, Wei Z, Cao Z. Automatic fault recognition for brake-shoe-key losing of freight train. *Optik*. 2015 Dec 1;126(23):4735–42.
11. Belongie S, Malik J, Puzicha J. Shape matching and object recognition using shape contexts. *IEEE Transactions on Pattern Analysis and Machine Intelligence*. 2002 Aug 7;24(4):509–22.
12. Cortes C, Vapnik V. Support-vector networks. *Machine Learning*. 1995 Sep 1;20(3):273–97.
13. Yegnanarayana B. Artificial neural networks for pattern recognition. *Sadhana*. 1994 Apr 1;19(2):189–238.
14. Krizhevsky A, Sutskever I, Hinton GE. Imagenet classification with deep convolutional neural networks. In *Advances in Neural Information Processing Systems*, MIT Press 2012 (pp. 1097–105). Cambridge, United States.
15. Rawat W, Wang Z. Deep convolutional neural networks for image classification: A comprehensive review. *Neural Computation*. 2017 Sep;29(9):2352–449.
16. Garcia-Garcia A, Gomez-Donoso F, Garcia-Rodriguez J, Orts-Escolano S, Cazorla M, Azorin-Lopez J. Pointnet: A 3d convolutional neural network for real-time object class recognition. In *2016 International Joint Conference on Neural Networks (IJCNN)*, 2016 Jul 24 (pp. 1578–1584). IEEE.
17. Yin X, Fan X, Wang J, Liu R, Wang Q. An automatic interaction method using part recognition based on deep network for augmented reality assembly guidance. In *International Design Engineering Technical Conferences and Computers and Information in Engineering Conference*, 2018 Aug 26 (Vol. 51739, p. V01BT02A018). American Society of Mechanical Engineers.
18. Ren S, He K, Girshick R, Sun J. Faster r-cnn: Towards real-time object detection with region proposal networks. In *Advances in Neural Information Processing Systems*, 2015 (pp. 91–99).

19. Dalal N, Triggs B. Histograms of oriented gradients for human detection. In *2005 IEEE Computer Society Conference on Computer Vision and Pattern Recognition (CVPR'05)*, 2005 Jun 20 (Vol. 1, pp. 886–893). IEEE.
20. Simonyan K and Zisserman A. Very deep convolutional networks for large-scale image recognition. *International Conference on Learning Representations*, 2015.
21. Nowak E, Jurie F, Triggs B. Sampling strategies for bag-of-features image classification. *European Conference on Computer Vision*, Springer, Berlin, Heidelberg, 2006 (pp. 490–503).
22. Bjerrum EJ. SMILES enumeration as data augmentation for neural network modeling of molecules. arXiv: Learning, 2017.
23. Flusser J, Suk T, Farokhi S, et al. Recognition of images degraded by Gaussian blur. *IEEE Transactions on Image Processing*. 2016;25(2):790–806.
24. Fitzgibbon AW, Pilu M, Fisher RB, et al. Direct least square fitting of ellipses. *IEEE Transactions on Pattern Analysis and Machine Intelligence*. 1999;21(5):476–80.
25. Gao Q, Yin D, Luo Q, et al. Minimum elastic bounding box algorithm for dimension detection of 3D objects: A case of airline baggage measurement. *IET Image Processing*. 2018;12(8):1313–21.
26. Hu MK. Visual pattern recognition by moment invariants. *IEEE Transactions on Information Theory*. 1962;8(1):179–87.
27. Simon L and Pawlak M. On the accuracy of Zernike moments for image analysis. *IEEE Transactions on Pattern Analysis and Machine Intelligence*. 1998;20(12):1358–64.
28. Suzuki S and Abe K. Topological structural analysis of digital binary images by border following. *Computer Vision, Graphics and Image Processing*. 1985;30(1):32–46.
29. Besl PJ and Mckay ND. A method for registration of 3D shapes. *IEEE Transactions on Pattern Analysis and Machine Intelligence*. 1992;14(2):239–56.

11 Combustion Timing Control of a Recompression HCCI Engine Using Negative Valve Overlap through Reverse Engineering

M. Suleman, Muftooh Ur Rehman Siddiqi, and Sundus Tariq
CECOS University of Emerging Science and Information

CONTENTS

11.1 INTRODUCTION

11.1.1 Background

With the advent of modern technology there is a major shift in focus of research toward renewable energy sources. The oil reserves are considered a major asset these days and research these days is focused on renewable energy systems. There is focus on production of making renewable fuels for engines and vehicles. All is being done to make efficient and environment friendly fuels by the help of which we can save the environment from the pollutants. There is need of an efficient combustion so that the fuel consumption can be minimized as well as the emissions from the burning of the fuel used. The fuels used in the internal combustion were changed but still there was need to develop an engine which has reduced emissions such as the NOx emissions and the hydrocarbon emissions so that the environment pollution level may be controlled. One such concept was to combine the characteristics of both the petrol engine and the diesel engine to get a more efficient engine which can produce low emissions. A study shows the environmental impact of pollution by transportation [1].

The toxic emissions from the vehicles may be the hydrocarbons along with carbon dioxide and carbon monoxide. These gases are harmful for the human body because the carbon monoxide is a poisonous gas while the carbon dioxide is responsible for producing greenhouse effect which is the major cause of increase in the overall temperature of the planet. The NOx emissions are also dangerous because they can chemically react with other gases and vapors to produce more deadly and poisonous gases. The NOx emissions are the main source of ozone formation which can have bad effects on the environment [2].

The ozone formation can be good for the upper atmosphere, but it can cause harmful effects when the human body is exposed to it. The ozone formation in the lower atmosphere may cause breathing problem for people and may be life threatening. Other emissions may be the particulate matter which is responsible for the production of smog and other pollutants which pollute the environment and thus may block the flow of air and cause the temperature of the atmosphere to rise. There is need for taking major steps towards improving the quality of the air so that the human beings breathe [1]. To get rid of the environmental hazards there is need to produce such an engine which can provide efficient combustion of fuel with minimum amount of NOx and other harmful emissions. With the growth in the economy and the increase in the number of vehicles and production from the automobile manufacturers it was needed to develop such an engine which has all these issues resolved. A step toward reduction of the emissions is the development of Homogenous Charge Compression Ignition (HCCI) engine [3].

11.1.2 Internal Combustion Engines

11.1.2.1 Otto Cycle

The working of the HCCI engine is based on the Otto cycle. The Otto cycle is a cycle for engine working which shows the behavior of the pressure inside the engine at the time of engine working. The Otto cycle shows the behavior of engine fuel in the engine cylinder. The Otto cycle shows idealized behavior of the engine fuel which consists of the behavior of engine pressure and volume during the movements of the engine piston. The Otto cycle gives excellent explanation of the energy production and transfer during the working of the internal combustion engine. The working of the engine can be explained by the ideal Otto cycle as well as by actual Otto cycle. The ideal Otto cycle shows the engine working during isentropic compression and expansion. The volume of the fuel remains constant while the fuel is being compressed in the engine. The isentropic process ensures that there is no heat transfer from the engine and the engine processes are adiabatic and reversible [4]. The ideal Otto cycle shows constant volume as the pressure is increased. The ideal Otto cycle provides with an explanation to the engine working. The actual Otto cycle gives a good idea about the actual engine performance. The actual Otto cycle shows the changes in volume and pressure as the engine piston moves up and down. As the intake occurs the volume increases with decrease in the engine pressure. As the compression stroke begins, the pressure starts increasing. The pressure reaches its peak value at the end of combustion [5]. At the end of combustion, the pressure starts to decrease while the volume starts increasing. The Otto cycle completes as the exhaust valve opens and the volume starts decreasing again and from intake stroke the next cycle begins.

11.1.2.2 Two-Stroke Engine

A two-stroke engine has a complete combustion cycle in two strokes of the engine. With only two movements of the crank shaft, the engine produces the ignition as well as the emissions from the exhaust chamber. The engine works as the movement of the piston from top dead center to bottom dead center; during this phase, the engine camshafts or other injection ports inject the fuel into the engine cylinder. The movement of the engine cylinder from bottom dead center to top dead center ensures the compression and ignition of the fuel in the engine cylinder. The injected fuel is then ignited by the spark plug or by compression, depending upon the type of the engine. The movement of engine piston from Top dead center (TDC) to Bottom dead center (BDC) starts the exhaust phase in which the exhaust gases are released from the engine.

11.1.2.3 Four-Stroke Engine

The four-stroke engine uses four strokes in the entire combustion cycle. These strokes are as follows.

11.1.2.3.1 Intake Stroke

Thee intake stroke begins from TDC and ends at BDC. During the intake stroke, as the name suggests, the fuel is taken in into the engine cylinder. The intake stroke ensures that the fuel is ready in the engine cylinder before the next stroke kicks in.

11.1.2.3.2 Compression Stroke

The compression stroke starts from the BDC and ends at TDC. The compression stroke is the main stroke during which the fuel injected into the engine cylinder during the previous stroke is compressed so that the ignition of the fuel is ensured. The compression stroke produces effort to compress the fuel up to the TDC. The compression of the fuel depends upon the type of fuel. Different fuels may have different behavior to compression by an engine piston.

11.1.2.3.3 Power Stroke

The power stroke starts from TDC and ends at BDC. During the power stroke the engine fuel compressed during the previous stroke is ignited. The ignition method is dependent on the type of the engine. The power stroke is the most important and powerful stroke which drives the engine by the crank shaft movement. The power of the burning fuel drives the engine based on the laws of thermodynamics.

11.1.2.3.4 Exhaust Stroke

Exhaust stroke is the last stroke during which the exhaust gases are emitted from the engine. It begins from BDC and ends at TDC. The movement of piston ensures that all the exhaust gases are emitted from the engine after which the intake cycle starts once again.

11.2.1 Homogenous Charge Compression Ignition Engine

The Homogenous Charge Ignition Compression Engine as the name suggests is a type of internal combustion engine which uses the characteristics of both the Spark ignition engine as well as the Compression Ignition Engine. The Homogenous charge means that the air fuel mixture is prepared before its entry into the engine cylinder. The mixture of air and fuel is then ignited using compression of the engine piston. The HCCI engine takes good characteristics from both the diesel engine and the petrol engine. The working of the petrol engine involves the use of a spark plug as the ignition trigger [6]. The fuel in the engine cylinder ignites and the engine is started. The air fuel mixture is introduced into the engine cylinder which is compressed and ignited using a spark plug. The diesel engine or compression ignition engine uses compression of the air for ignition. The air is compressed in the engine cylinder and the fuel is sprayed which results in ignition in the engine cylinder. The ignition takes because the condition after compression of air in the engine cylinder is very much favorable for combustion [7]. The high pressure after compression and the highly inflammable fuel in the engine cause the ignition. The good attributes taken from both the engine make up the HCCI engine. In the HCCI engine, the fuel and air are mixed as in the Spark ignition (SI) engine case, but the ignition of this fuel takes place by compression as in the Compression ignition (CI) engine case.

This is a challenging task because the ignition is not directly controlled instead it depends on the cylinder pressure, temperature, and the fuels used. The ignition of fuel in HCCI engine is completely dependent upon the type of fuel and the engine environment. The HCCI engine has more efficient combustion as compared to both SI and CI engine because the engine has the advantages of both the CI and SI engines. The advantages from SI engine are that the air-fuel mixture is premixed. The premixing of air-fuel ratio ensures that the fuel mixture is in its best condition for ignition, i.e., it is homogenously mixed. The homogenous mixture is then introduced into the engine cylinder where it is ignited by the spark plug. The advantage of using premixed air fuel mixture is that it has relatively lesser emissions known as NOx emissions compared to CI engine [8].

The lean mixture of air and fuel ensures that the combustion is clean with very less amount of NOx emissions. The compression of fuel is done up to a certain limit because the fuel and air together can only be compressed to a certain point. When the air-fuel mixture is compressed further, the engine produces a knocking effect. The SI engine has the property that the air and the fuel are premixed [9]. Now this means that the compression ratio cannot be kept up to very high values. If the compression is forced on an SI engine the unburned fuel in the engine which is a waste may start to auto-ignite. The auto-ignition of this residual fuel causes a knocking effect in the engine. The other disadvantage is that when the engine fuel is compressed to a very high ratio, the ignition may have negative impact on the engine parts. It may cause damage to the spark plug ignition system which is highly undesirable. On the other hand, the CI engine has high compression ratios as compared to SI engines. The main reason for this is that the air is compressed in CI engine and then the fuel is ignited when it is sprayed [10].

The compression of only air is easy as compared to the compression of an air fuel mixture. The mixing of the fuel with compressed air in the engine cylinder does not guarantee complete combustion of the fuel and thus the fuel burned has a lot of emissions and particulate matter. Although the emissions are high in the CI engine, the efficiency of the CI engine is also high as compared to the SI engine. The combustion of fuel with this process ensures that a lean mixture is used. The use of lean mixture is always very advantageous because the combustion takes place at a low temperature. When the combustion occurs at a low temperature it means that the emissions will be very less from the engine fuel [11].

The good attributes of the spark ignition engine are the premixed air fuel mixture. The premixed air-fuel mixture ensures that the emissions from combustion are minimum. The homogenous charge ensures that the clean combustion of fuel occurs inside the engine cylinder. The compression of the fuel is an attribute taken from the diesel engine. It ensures that the compression ratio of the fuel is high, and the air-fuel mixture is lean. These attributes from both the engines result in high efficiency of

HCCI engine as compared to the conventional SI engine. The increased combustion efficiency along with the reduced losses is the reason the HCCI engine is a step ahead of the SI and CI engines [12].

11.2.1.1 HCCI Engine Challenges

The functioning of an HCCI engine may seem very easy but it is not that simple. The main challenge in the implementation of HCCI engine is the combustion control of the engine. The combustion or ignition control as it may be called is the most important part of the engine working cycle. The combustion phasing determines when the ignition will occur in the engine and when will the chemical reactions occur which are responsible for the fuel combustion in the engine. The main research focus toward HCCI engine has been toward the combustion control of the engine [13]. The difficulty lies in the fact that the engine has no component for carrying out direct ignition. The engine has the fuel and the compression which are used for the ignition. The absence of the spark plug makes the ignition quite a challenging task. The combustion primarily depends upon the composition of the fuel mixture, the temperature of the gases inside the engine cylinder, the pressure, and some other factors. The combustion control is the main part where the engine will carry out the combustion in its most efficient form. When the combustion is not properly controlled, the engine may not work that efficiently. The combustion should be carried out at the most efficient point without which the efficiency of the engine may decline rapidly, mainly because the engine will not be able to perform efficiently until the combustion is not carried out efficiently [14]. The combustion should be carried out at the exact range in which the HCCI operation is ensured, outside that range the HCCI operation may not even initiate because the engine will not be able to perform correctly. The early ignition has its own problems because this may lead to knocking problems in the engine. When the ignition is too late, it may lead to improper burning of the fuel which wastes fuel and releases large amounts of emissions into the environment. The HCCI implementation is carried out by different temperature and pressure techniques which also provide the basis for HCCI engine modelling [15].

11.2.1.2 HCCI Engine Combustion Control

This work presents the combustion timing control of an HCCI engine. The control of the engine combustion phasing is very important since it will help in determining the performance of the engine over different sets of combustion conditions. The combustion phasing in the engine is affected by different parameters which include the temperature and pressure as well as the fuel used in the HCCI engine [16]. There are various methods which influence the temperature and pressure inside the HCCI engine. The control techniques used can be the control through variable valve timing (VVT) or by introducing fuels of different proportions which can influence the combustion timing of the HCCI engine. One of the techniques used to control the HCCI engines is the negative valve overlap (NVO). The NVO is a technique in which a portion of the exhaust gases is trapped in the HCCI engines which can lead to the control of the combustion phasing in the HCCI engines [17]. The influence of trapped exhaust gases in the recompression HCCI engine is such that it can influence the in-cylinder temperature and pressure. The changing in the engine environmental conditions helps in the control of the combustion phasing of the engine. Through reverse engineering, an HCCI engine model will be checked for its open-loop characteristics and then the control of the combustion will be done using NVO.

11.3 HCCI ENGINE MODELLING FROM LITERATURE

The modelling of the HCCI engine is a very important step towards simulation of the engine. The mathematical modelling calculates all the engine parameters and evaluates them to verify the proper working of the engine. The laws of thermodynamics help to provide enough material for the successful modelling of the HCCI engine [18]. To successfully control the engine, the engine modelling is very important to which the controlling techniques or the optimization algorithms are applied. Although an exact accurate model of the engine is nearly impossible to design but by approximating some parameters and by keeping in mind the main control features of the engine, the engine modelling can be completed. The derivation of these models helps us in collecting the main engine parameters which allow us to successfully model the engine and run the simulation for its analysis [19]. Several models were created for the HCCI engine. These models were developed over the years and each model provided its own set of parameters and setups for the analysis of the HCCI engine. The engine models were developed, and different control techniques were applied to each model.

The models developed for the control of the HCCI engine were Single input single output (SISO) models initially. These models used a simple approach to model the HCCI engine [20]. A similar approach to that of closed-loop control was observed. The control technique used was the variable valve actuation. The variable valve actuation is pretty helpful in determining the combustion control and thus provides a good solution to perform the control of the engine [21]. In another control technique the control was achieved by observing the combustion in the engine with the help of sensors and microphones. The data observed provide the detail of the engine combustion along with knocking phenomenon if any observed. The information of engine knocking and other combustion behavior is very important because it gives a great deal about the engine working [22].

The Multiple input multiple outputs (MIMO) systems have also been designed in the literature. The literature shows a control system using two PID controllers which work alternatively. The function of the PID controllers is to manage the combustion control. The control over the trapped residuals and the IVC (inlet valve control) is the main technique for carrying out the control of the combustion. The actuating valve for the IVC is responsible for changing the compression ratio. The change in the compression ratio allows for the optimum selection of the compression ratio for the device which can be used to achieve the best and optimum engine performance [23].

The literature shows development of MIMO systems in which each cycle of the HCCI engine is controlled. The engine has multiple cylinders in which the combustion occurs. The multicylinder control of the engine is very important since it will show the control on a much larger scale. The multicylinder control of the engine ensures that the control technique applied is more robust and that it ensures the combustion in the desired range [24]. The cycle-to-cycle control of the engine helps in developing a control system which is more capable of providing the necessary parameters for the control of the engine [25].

A control for HCCI engine was developed fueling at gasoline. The MIMO model was used in the control system [26]. Model predictive control (MPC) was used to control the engine performance. The use of MPC provides a good solution for the working and control of the HCCI engine. The control technique was developed for the engine fuel injection and the variable valve actuation. These two are the most important and commonly used methods for increasing the engine efficiency and the control time [27]. Optimal control techniques are very important since they use the optimization of the main engine parameters and the control techniques which are demonstrated in Ref. [28]. By applying constraints to the engines working, the working may be observed in different conditions and under different circumstances. Another MPC of the engine is shown in Ref. [29]. The control is done by controlling the nonlinear model of the engine which requires simultaneous evaluation of the engine parameters which makes it quite a challenging task. In Ref. [30], a MIMO system has been implemented which represents two inputs and a single output. Split compression technique was used for the control of the combustion. Variable valve actuation and exhaust gas recirculation (EGR) techniques were used for the control of the HCCI engine. The control model uses a double input vs a single output technique for the combustion control of the engine [31]. In Ref. [32], the solution to a highly nonlinear model of the engine was provided by a linearization technique which provides a tracking technique to control the engine combustion. The extremum seeking technique is used to determine the parameters of the engine which will be used to facilitate the control of the HCCI engine [33].

A physics-based model of the HCCI engine was developed by Shaver and Gerdes [25]. This work shows that a low-order model can be developed which is simple and which can be described with a very few states. The advantage of this is that the model is easy to understand and that the parameters make it simple to design a controller or observer for the system. A model based on the rebreathing and recirculation in the engine was developed in Ref. [34]. Rebreathing and recirculation are the phenomena which are very efficient for the engine combustion. The process of rebreathing and recirculation uses a simple approach to introduce the exhaust gases in the engine which then gives the desired range of combustion for the engine. A simple eight-state model was developed in Ref. [35]. The purpose of this model was to study the misfiring in the HCCI engine. Misfiring is highly undesirable because a proper combustion is not ensured due to misfire which is the main challenge in engine combustion. As the combustion occurs in the engine cylinder, there is always an interaction with the engine cylinder walls. These combustion interactions are considered in a model developed in Ref. [36]. The conservation and thermodynamics laws were used to develop a model.

The effect of recompression in comparison with rebreathing was observed in Ref. [37]. The engine environment was studied in 1-D which then made it possible to collect and carry out simulation. The use of vapors of fuel was done in Ref. [38]. It was observed that as the fuel is injected as vapors into the engine cylinder, the alkaline reactions in the fuel forces it to change its properties. These changing properties have a significant effect on the engine performance. The engine's combustion properties are altered, and the ignition timing of the engine can be controlled when the engine vapor's properties are changed to a certain level. A study on the latest trends and control techniques in the HCCI engine was done in Ref. [39]. The paper shows the development in the HCCI engine field and the properties of the engine as well as the control of emissions in the HCCI engine. It shows the environment friendly techniques for the engines which will reduce the emissions and contribute toward the goodness of the environment.

Intake charge temperature and the EGR have a very important role in controlling the engine's combustion. The combustion of engine using these techniques is helpful in determining the fuel properties. In this study, a biodiesel-fueled HCCI engine was studied and the effect of the intake temperature and the EGR was observed on the engine [40]. An ozone generator was used at the intake to control the combustion phasing of the HCCI engine [41]. The intake was provided with oxygen and air. The effect of these two gases was checked on the combustion phasing of the HCCI engine. The study shows that the important engine parameters such as the indicated mean effective pressure (IMEP) and the emissions from the engine can be controlled by varying the concentration and amount of ozone from the ozone generator.

The use of ozone is a very important technique for combustion control of the HCCI engines. The ozone is used as an oxidizing agent which helps in controlling the combustion of the HCCI engine. With the development of modern technology, the ozone generators have become small and compact. This is an advantage since it can be used in different vehicles and engines. The combustion control of the HCCI engine is done using ozone gas at the intake [42]. The thermodynamic properties of the HCCI engine fuel are varied by changing the composition of fuel in the HCCI engines. The fuel used can be sprayed which have a very important effect on the combustion phasing of the HCCI engine [43].

The effect of a low-octane number fuel in HCCI engines was observed, and the study shows that by using a low-octane number, the HCCI engine cannot induce the best combustion. The knocking problem is more enhanced when the fuel used is of low-octane number [44]. The spark-assisted HCCI engine was used along with the technique of after treatment of the exhaust gases. The study shows that the spark-assisted HCCI engine along with after treatment can significantly reduce the emissions from the engine which improves the efficiency of the engine [44]. The performance map of an HCCI engine was developed which was fueled on ethanol. The important engine parameters such as engine load, engine speed, intake temperature, and others were observed [46].

11.3.1 Charge Compression Ignition (CCI) Engine Combustion Control Techniques

11.3.1.1 CCI Engine Modelling for Control

The HCCI engine control may not be that simple, but the control of the engine depends upon the general engine factors which should be kept in mind. These factors include the general engine functioning parameters. These include the intake air of the engine. The engine requires air intake for the combustion of the fuel in the cylinder. The slightest changes in the air intake temperature have a significant effect on the chemical reactions in the engine cylinder. This control method relies on the change in temperature of the intake air for the engine. The intake air temperature may be changed by several methods which provide its use in a better combustion process and ensures the efficient combustion of the fuel in the engine cylinder. The fast-thermal process is also a major technique in changing the intake temperature of the HCCI engine [47]. The variation in intake temperature for a multicylinder engine is also manageable [48]. This includes the control of all the engine thermodynamics which results in the control of the intake temperature of the engine [49]. The change in the intake temperature for a gasoline engine is also done [50]. By changing the input air temperature, the combustion process is affected because these processes are involved in changing the chemistry of the chemical reactions. Different the levels of heating, different will be the combustion results [51]. Ethanol was used as a fuel in the HCCI engine and the compression ratio of the engine was fixed as well as the engine speed. The results showed that keeping the temperatures in a certain range low temperature combustion can be achieved with very low level of emissions [52]. The results of variation in intake temperature in an n-butane-fueled HCCI engine are also done. The compressions ratio is kept constant as well as the equivalence ratio [53].

11.3.1.2 Combustion Control by Variable Compression Ratio

The compression ratio of the engine is the ratio of maximum volume to the minimum volume in the engine cylinder. Usually a high compression ratio is considered as ideal because it ensures that the efficiency of the engine is higher. The higher compression ratio ensures that the minimum fuel is used, and the output is maximum. The compression ratio of an engine has a great effect on the combustion of the engine since it determines the amount to which the fuel will be compressed for ignition. A variable compression ratio engine is presented which shows the control of ignition with respect to the change in the inlet air temperature [54].

The variable compression ratio is also evaluated for a multicylinder engine which shows the combustion control due to the change in the proportion in which the air and fuel are compressed. A closed-loop system is simulated which evaluates the variable compression ratio along with the speed and load of the engine [55]. The lesser the intake temperature, the more can be the compression ratio. This behavior was studied and simulated, and the results were shown [56]. The compression ratio was increased, and the effect was observed by changing various intake temperatures. The maximum compression ratio translates in the form of the pressure or IMEP developed in the engine cylinders. The results show that at the lowest intake temperature, the compression ratio was the highest. The intake temperature was kept constant and the compression ratio was set to different values. It is observed that the engines' equivalence ratio and the engine speed have a large range of operation when the compression ratio is set to the highest value [50].

11.3.1.3 Combustion Control by EGR

EGR is another important technique for controlling the combustion process in the HCCI engine. The EGR is used to change the temperature of the air inside the cylinder. EGR is also helpful in cases where a chemical change in the air mixture in the engine cylinder is desired. The three main control parameters for the control of HCCI engine are the in-cylinder temperature, the pressure, and the temperature. The EGR can have a significant effect on all three of them since EGR can produce several effects on the engine cylinder. The EGR can be used for the dilution of the gas mixture in the engine cylinder. It can also be used for the chemical changes desired in the combustion process. The thermal effects of EGR are also very important since the exhaust gas has a very important effect on the thermal behavior of the engine.

The hot gas in the engine exhaust can be very important in changing the engine thermodynamics. The chemical reactants of the HCCI engine are the gases that are produced by the engine exhaust. One of the very common gases in the engine exhausts which exhibit the combustion are the CO and the NO. The CO is the most important since it is a combustible gas and it has a very important role in changing the characteristics of the HCCI engine combustion. The CO can affect the engine pressure whereas the NO concentration may speed up or slow down the chemical process in the engine. The NO gas and formaldehyde in the engine cylinder may delay the ignition process [57]. A very important advantage of using EGR in HCCI engines is the use for increasing the in-cylinder temperature of the engine. The cooling effect in the engine may slow down the combustion process and it may occur at a high temperature which will increase the emissions from the engine. The concept of the EGR provides high intake air temperature which further goes on to carry out the combustion at a comparatively low temperature in the engine cylinder. The advantage of low-temperature combustion is that the NOx emissions are significantly reduced which is healthy for the environment and increases the engine efficiency [57].

One of the methods to ensure that the engine combustion is facilitated by the EGR is the concept of NVO. This technique is based on the EGR but uses a different approach for the introduction of the exhaust gases into the engine cylinder. During NVO the exhaust valve is closed a bit earlier as compared to the other engine operation. This traps some of the residual gases of the engine into the cylinder which then functions as EGR for the HCCI engine [58]. The direct injection of fuel into the engine cylinder will then make sure that the combustion process carries on and the engine works properly. The chemical process after NVO and the direct injection of fuel into the engine cylinder produce the hydrogen gas through some chemical mechanism which further aids in the combustion process [59].

11.3.1.4 Combustion control by VVT

VVT is another very important method to increase the efficiency of engine combustion process. The VVT is the technique in which the inlet and the exhaust valves of the engine are controlled based on the performance of the engine. The VVT ensures that the most is collected from the engine running under certain conditions. The VVT controls the engine camshafts which regulate the use of the fuel injection and the exhaust valve opening. The opening and closing of these valves if done effectively is very helpful in fuel saving. The valve opening is usually controlled by the camshafts which are connected to the crankshaft. The camshafts open and close the inlet and exhaust valves so that the injection of fuel into the cylinder and the emission of exhaust can be done properly. The intelligent control of these valves is the key which is to determine the valve timing with changing engine speeds and the complete or incomplete combustion of the fuel [60]. The operating range of the HCCI engine was determined with built-in turbo charge and the VVT technology along with direct injection of the fuel [61]. The simulation environment for the VVT HCCI engine is very important since it helps in carrying out engine tests and evaluations with minimum effort and accurate results [21].

11.3.1.5 Combustion Control by NVO

NVO is another technique for controlling the combustion phasing in the HCCI engine. The NVO concept comes from trapping the residual gases inside the engine cylinder which change the operating conditions and the environment in the engine cylinder. The NVO changes the basic operating conditions of the engine which provides the user to control the combustion of the engine. The effect of fuel composition changes was studied in an HCCI engine with NVO. The NVO technique was used to control the ignition of the engine. The composition of different fuels was varied to determine its effect on the performance of the engine. It was observed that the changes in the fuel composition can be very useful to observe the changes in the performance of the HCCI engine. The sensitivity of combustion at a high temperature to the fuel consumption is observed and it is noted that the fuel composition is very less sensitive to the high temperature combustion in the HCCI engine [62].

Ion current sensing is a technique which can be used to monitor combustion process happening inside the engine cylinder. The approximation of the combustion process inside the engine cylinder can be very helpful in determining the combustion process occurring inside the engine cylinder. The ion current sensing technique was used along with NVO and the results show that the combustion phasing signals can be more defined with the NVO technique. The signals generated for the combustion phasing with ion current sensing are more accurate at lower temperature and low load on the HCCI engine [63]. Direct fuel injection was used along with the NVO technique to observe the combustion behavior of the engine. The observation showed that the combustion of the engine is highly affected by the NVO technique which changes the temperature and the environmental conditions inside the engine cylinder and provide a good control over the engine operating range [64]. Partially premixed combustion technique is used in the engines along with the NVO.

The partial premixed mode uses fuel in which a premixed fuel is used prior to combustion. The low loading operation of the engine is observed during this mode [65].

The effects of NVO on the emissions in an HCCI engine are shown in Ref. [66]. The paper shows the experimentation done on the HCCI gasoline engine using the NVO. The results show that the emissions from an HCCI engine are reduced significantly when the NVO technique is used. Different valve settings were used to determine the characteristics of the gasoline engine used in the combustion process. The output was observed as the engine workload on the engine. Combustion control in an HCCI engine is a very challenging task which can greatly change the performance of the engine. The combustion control of an HCCI engine was studied using NVO. A two-stage direct ignition was used in this work which showed that this is a promising technique for controlling the combustion in the HCCI engine [59]. A study was done on the behavior of cycle temperatures along with the NVO to determine the behavior of the engine [67]. The establishment of understanding of the behavior of the engine cylinder combustion phasing along with the NVO is very important since it helps in evaluating the performance of the engine over different ranges. The various ranges of NVO operation were applied to determine the temperature of the in-cylinder combustion process. Auto-ignition is a very important property of the HCCI engine. The auto-ignition of the fuel occurs when certain criteria have been fulfilled for the combustion in the engine cylinder. In Ref. [68] a mixture of ethanol/air mixture was used in the HCCI engines working in the NVO mode. The main purpose of this study was to determine the heat flow inside the engine cylinder along with the engine dynamics which help in the determination of the engine working characteristics.

11.4 HCCI ENGINE MODELLING AND CONTROL DESIGN

The modelling of a single-cylinder HCCI recompression is presented in this chapter to show the working of the engine. The HCCI engine is modelled through reverse engineering and is then linearized around some operating conditions to get a controllable model of the recompression HCCI engine.

11.4.1 Modelling of HCCI Engine

The model used for controlling the combustion phasing in this work is taken from Ref. [69]. The modelling of the HCCI engine is very important since it determines the control approaches that can be implemented toward the model of the engine. The modelling of the engine should correctly represent the operating behavior and conditions of the engine. It is very important that the model has a close approximation with the actual engine because all the control techniques that will be implemented will be based upon the behavior of the model to certain sets of inputs. The modelling should be done such that the outputs and the inputs are separately defined and can be controlled with respect to each other. To ensure that the modelling is validated, reverse engineering is done to obtain a model of the HCCI engine. The equations obtained through reverse engineering are transformed into matrices which help a lot in applying a linear controller to the model obtained.

The HCCI engine mainly has two outputs which can be changed by different techniques. The outputs are the work output and the combustion angle of the engine which determine the performance of the engine. These outputs show how the engine has performed over the inputs that have been applied to them. The observation of these outputs also shows the combustion process in detail inside the engine. These outputs are dependent on the engine temperature, pressure, and the other conditions inside the engine cylinder. The model developed has the capability to control these outputs depending on the input. The combustion and the work output of the engine are represented by a set of equations which is linearized around a set of operating conditions for the engine. The states of the engine are not directly measurable from the engines' actual model so an observer is used which will correctly estimate the states that influence the performance of the engine. The controller is developed for a single cylinder model of the engine on which the model is tested for the different inputs along with the disturbances that can occur during the working of the engine. Although the controller developed is a linearized model it can operate on several different operating ranges of the engine. The linearizing around a single point does not narrow down the operating range of the model. The model can still work on different sets of inputs and controlled instructions. The model is developed such that it can control the unstable states of the engine over different sets of input instructions. The model is developed such that controllers can be implemented on them to control the outputs of the engine. The control of the engine is implemented by states which are highly dependent on the outputs and inputs of the engine [70]. The linearized model of the engine is shown in Figure 11.1.

The model shows the matrices for the state space model of the system. The matrices show that the linearized engine model is linearized around the operating points shown in Table 11.1. The model is linearized around the operating point so that we can get a linearized engine model which can work under different controllers and can provide the output for any controller that is applied to it.

$$A = \begin{bmatrix} 0.43 & -1.16 & -0.27 & -0.11 \\ -0.03 & 0.25 & 0.13 & -0.37 \\ 0 & 0 & 0 & 0 \\ 0 & 0 & 0 & 0 \end{bmatrix}$$

$$B = \begin{bmatrix} 0.02 & 1.90 & -0.92 \\ -0.02 & -0.18 & 0.45 \\ 1 & 0 & 0 \\ 0 & 1 & 0 \end{bmatrix}$$

$$C = \begin{bmatrix} -0.04 & -0.34 & -0.0063 & 0 \end{bmatrix}$$

FIGURE 11.1 Linearized HCCI engine model.

TABLE 11.1
Operating Point at Which the HCCI Engine Control Model Is Linearized

Parameter	Value	Units
Engine speed	1800	rpm
IVO	65	CAD
IVC	205	CAD
EVO	480	CAD
EVC	640	CAD
Mass of fuel injected per cycle	10	mg
NMEP	2.5	bar
CA50	363	CAD

11.4.1.1 States of the Model

The model designed is used to control the combustion phasing and the IMEP of the engine. The states that influence these outputs are the mixture temperature, fuel concentration, concentration of oxygen, and the cylinder volume at the closing of intake valve.

11.4.1.2 Recompression in HCCI Engine Model

Recompression in an HCCI engine is a phenomenon which ensures that the temperature inside the engine's cylinder is brought about to some desired level which will contribute to the lowering of the emissions from the engine as well as the losses in an HCCI engine. During recompression both the inlet and the exhaust valves of the engine are closed. The recompression is a cycle which utilized the energy in the fuel to determine the conditions of the engine after the recompression cycle. The main influence of the recompression cycle comes from the injection of fuel at the end of the recompression cycle. During recompression, since both the valves are closed it provides an environment in which the air is compressed at leaves behind the conditions in which advantage can be taken from the injection of fuel after the recompression cycle. After the injection of fuel, a process of heat transfer occurs in which heat is exchanged between the engine cylinder and the fuel due to evaporation of the fuel. The heat exchange process follows the first law of thermodynamics in which the transfer of heat occurs. The transfer of heat that occurs during the recompression process is nonlinear. The time duration of the recompression is dependent upon the valve timings and the speed of the engine.

11.4.2 Proposed Controller

The proposed controller is a PID controller which can be tuned by using genetic algorithm (GA). The PID controller is a control mechanism which uses the proportional, integral, and the derivative gains to control any process which has the error generating from it. The PID controller constantly generates an error signal between any set point value and the control variable which is then minimized using the gains of the controller.

The diagram above shows the block diagram which is the schematic of a typical PID controller. The figure clearly indicates that the proportional, integral, and derivative control parameters are contributing together to provide an input which will eliminate the error between the parameters. The error is generated by the difference between some reference value and some control variable which needs to be controlled. The calculation of error and its elimination at every point is the basic task of the PID controller and thus it gives a desired value at the output by eliminating these errors.

Each of the terms in the PID controller carries its own significance in the elimination of error from the output. The term proportional in the PID controller is directly proportional to the error that is generated between the set point and the variable that is at the output. If the error is increased between the two points the gain from the proportional controller is increased as well and it will try to compensate the error that is generated.

The I term in the PID controller has its own significance since it uses the integral control for the elimination of error from the output. The term integral compensates for the remaining error in the system and keeps a history of the system error that is generated. The integral term then eliminates this error and the system's performance is improved.

The derivative portion of the controller makes an estimate of the error that will be appearing between the set point and the variable at the output. The derivative term in the controller compensates for the error by taking a measure of the value of the controller which is changing with respect to time. By computing the rate of change of error, the error is eliminated at the output and the user gets an error-free output [70].

With all these parameters explained, there is a challenging task of tuning these PID parameters to the performance of the system. The right combination of these parameters is very important since it will determine the performance of the system. The PID control parameters are needed to be in control in accordance with each other and keeping in mind the performance of the actual system. By considering all these parameters it becomes easy to predict accurate values of these parameters. These parameters are derived for each of the system and can only be tuned once the performance of the system is in the stable range. The response of the system can be checked for some initial values of the PID parameters, but these values are needed to be changed for the specific system. The parameters are then tuned by refining these parameters which is accomplished by an introduction of a bump or disturbance in the process which will ensure that the controller is working properly without any problems and that the controller returns to the main set point even after the introduction of disturbances in the system. By the successful tuning of these parameters, these can be used to define problem-specific control action which will be used to bring the output of the system to the desired value. The control action is the output of the controller which will be used to bring back the output to the desired point from the error that was generated in the system. The main working of the PID is dependent on the tuning of the P, I, and D parameters [71].

There are various techniques for the tuning of the P, I, and D parameters which are developed over the years. Each of the techniques has its own advantages and disadvantages depending on the system they are being applied to. One of the basic techniques in the determination of the PID parameters is the Z-N method or Ziegler-Nichols method. This method uses the simple calculation of the PID parameters by using some basic knowledge of mathematics. By considering one of the parameters to zero, the others are estimated which will then be assumed to be zero to calculate the other variables. The Z-N method uses a preset table with defined values which will be used to calculate the values of the PID parameters. Other similar techniques were also developed over the years which were like the Z-N tuning method. These techniques were the Cohen-Coon parameters technique and the Relay method which is also known as the Astrom–Hagglund method. These methods use the empirical techniques for the evaluation of the PID parameters which has its own advantages and disadvantages. The advantage of the tuning using these techniques is that these methods are fast and can provide a very quick solution to the tuning of the PID parameters. These techniques, however, have their own disadvantages which are needed to be considered in order to successfully evaluate the performance of the system. These parameters have little or no knowledge about the changes that might occur in the system due to certain nonlinearities and unexpected changes in the system. These techniques only have a set of PID parameters which is used to control the system. Any robustness in the system's performance and error tracking is absent because these techniques are not intelligent enough to cope for the unexpected changes in the system [72].

There are solutions to these problems in the form of heuristic adaptive algorithms which are used to tune the parameters of the controller in such a way that the PID parameters are tuned intelligently and that there is no deviation from the output set point in the performance of the controller. These artificial intelligence algorithms have their own advantages which allow them to opt for the unexpected changes in the system and the user can easily evaluate the performance of the system by these adaptive algorithms. There are many techniques for the evaluation of the PID parameters. The heuristic algorithms are artificial intelligence algorithms which evaluate the performance of the system and then provide the tuned values of the system which can be used to evaluate the performance of the system over a range of changes in the set point of the system [73]. One of such algorithms is the GA which will be used for the tuning of the PID parameters in case of the HCCI engine model.

11.4.3 GA-Based PID Tuning

GA is the optimization algorithm which uses the method of natural selection among living species. This optimization technique focuses on the reproduction processes of living organisms to provide an optimization solution. The GA focuses on a population of species to get the desired solution. It works on the principle of natural selection which runs tests on the population, and we get the best possible outcome from the population of species. GA uses the definition of mutation and genetic crossover of the chromosomes to get the desired set of best species. The algorithm works based on the best selection of species from a population. The population has several species which are evaluated based on the value of their objective function. The objective function for each of the species will determine the quality of that specie. From a population of species all the species are evolved toward betterment. The parameters for each of the solution can be changed and are passed in their best form to the next generation. This algorithm works on the betterment of the solutions with each iteration. In an iteration, the best available species are chosen and are passed on to the next generation. The next generation then evaluates the best solutions out of the given species and the process continues. The iterations are carried up to the point at which the maximum iteration level is achieved or the species in the generation have an optimal solution. To implement the GA to an optimization problem, the problem must be present in the genetic form to apply the parameters of genetic mutation and crossover. The next most important requirement is the fitness function of the optimization problem which determines the quality and selection of a particular specie. The algorithm works based on the evaluation of the species in a population. The initial population is selected, and the algorithm starts its iterations. The fitness value of each solution is analyzed and compared and the best solutions are advanced to the next generation where the process continues [74]. The initialization of the population is usually done manually. The population is evaluated for the best possible solutions. The number of population solutions may vary depending on the type of optimization problem. The solution space usually consists of hundreds and thousands of solutions in each population; each of them are filtered based on the fitness function evaluation and are passed on to the next generation. The possibility of many solutions gives a perfect opportunity to get the global maximum from the optimization problem. The more the solutions, the more will be the available space for the solutions to exist. For evaluation of the best solutions, the population solutions are often placed in the desired area which also helps in evaluation of the global optimum from the set of solutions. A flowchart of the GA is shown in Figure 3.3. The selection of a population is based on best available solution of the optimization problem. The lesser effective solutions are eliminated from the solution space by the action of the fitness function. The fitness function determines the value for each of the solution. The algorithm automatically selects the best available solutions from the possible available. There might be options available for the selection of the most suited solutions but in most cases the newly generated population is selected from the initial population by the action of the iterations. The fitness function plays a very important role in determination of the next generation of species from the initial population. The fitness function may be derived from the system properties. The system behavior gives an indication of the fitness function and its behavior.

The selection of next generation is based on the reproduction process of species. The generation of a new population takes place in the exact same manner. Each solution in the new generation is made from asset of two solutions in the initial population. The new child solution is generated from two best parent solutions to get closer toward optimization. The newly generated child solution then gets involved in iterations to produce further solutions for the optimization problem. The crossover takes place in the same manner as in the biological species and the newly generated child solution has good characteristics from both the parents. The iterations for the GA are not infinite, instead they terminate when a certain limit is reached.

The termination of the optimization and the generation of new populations stop when the most optimum solution is achieved. There may also be a halt in the iterations due to the constraints applied to the iteration process. The number of generations can be selected which helps in maintaining a certain level of restrictions for the optimization process. The GA-based PID tuning is a very important application of the GA which provides an excellent solution to the tuning of the PID parameters for the control purposes. The PID parameters are controlled according to the problem and the output is set to the desired value which gives a very good value for the parameter tuning of the PID controller. The PID tuned with the help of the GA has a very robust behavior as compared to other tuning parameters which is very important for the functioning of the PID controller [75].

11.4.3.1 Limitations of GA

GA is a complex method of optimization which involves the evolution of species from one generation to the next. The GA may not take much for simple simulation of fitness functions, but in case there are other complex problems which have more complex fitness functions may not be evaluated that easily by GA. Dealing with hundreds and thousands of solutions for a complex fitness function may not be easy to evaluate since it will require immense computing power as well as more time for providing the solution of the optimization problem. Evaluation of a complex optimization problem may prove costly which might not be favorable keeping in mind the budget and time. Another problem with GA is that it has the tendency of getting stuck on the local optimum instead of searching for global optimum. The problems which have several local maxima may not be easy to optimize because of the local maxima the solution may not be easy to find. The complexity may increase when performing optimization on a complex system. The GA performs mutation and crossover to get the new species and provides a whole generation of solutions. The problems in which decision-making is involved may not be easy to solve using the GA. The GA may not be able to provide accurate results for random search spaces. The speed of GA may not be as fast as other search algorithms. The GA may take more computation time because of the nature of the algorithm itself. The instruction set for the algorithm makes a new generation of species from the parent species and it may be very time consuming as compared to other search algorithms [76].

11.4.3.2 GA Literature

To improve the performance of the permanent magnet synchronous motor the traditional PI controller technique was improved by the use of multiobjective GA [77]. The application of GA is very important for such multiobjective problems since it uses all the required parameters of the motor and hence the optimization can be performed efficiently. The GA optimization was applied to the rod pumping well. The purpose of its application was to determine the speed control model for the rod pump which proves quite advantageous in the production of oil [78]. A chemical kinetic model of the HCCI engine was optimized using GA. The target was a model of HCCI engine which was fueled with n-heptane. The GA was applied to the data obtained from actual engine and the engine performance was observed by mixing the fuel with the reformer gas [79]. The performance of a gasoline fueled engine was optimized by using GA. The output of the engine is determined by the torque and speed. The engine and full load torque of the engine were optimized by the use of GA [80]. The method of weighted sums was used to enhance the engines performance. Camshafts control the fuel flow into the engine cylinders. The camshafts of the engine play a major role in engines performance by providing the necessary pressure and inertia so that the desired amount of fuel enters the engine cylinder. GA was applied to the objective function derived [81]. The engine intake, valve timing, and the exhaust control were optimized by using the GA. The optimization was carried out using single and multiobjective algorithms. The development of engine model gave an accurate solution for changing the further engine parameters. The optimization provided important information about an improved engine design which will lead to the better performance of the engine [82]. The vehicle active noise was controlled and optimized using GA. The transfer function for the engine sound was modelled and the GA was applied [83]. The intake port has a very important role in engines performance since it determines the amount of fuel, the power absorbed from the fuel, and the amount of exhaust gases that are emitted from the combustion. By accurately optimizing the parameters of the intake port of the engine, the engines performance may be enhanced [84].

The parameters of a small diesel engine were optimized by GA. The multiobjective GA was used to carry out the optimization process. The approximation of parameters provides a good insight into the engines performance which can be used to determine the engine characteristics. The optimized engine parameters can be used for the engine control [85]. The emissions from the engine are very problematic and are always needed to be controlled. The multiobjective GA was used for parameter estimation and eventually resulted in the reduction of emissions from a diesel engine [86]. A kinetics-based model of the HCCI engine was optimized by GA. The optimization process helped in carrying out the parameter estimation of the kinetic model which then was used to develop a multiobjective optimization scheme for the engine [87].

11.5 SIMULATION RESULTS AND ANALYSIS

This section represents the simulation results for the proposed controller of the HCCI engine. The proposed controller was a PID controller which as tuned using GA. The usage of GA is of prime importance since it provides the user with optimized results for the proposed PID controller. The results were compared to the proprietary algorithm by MATLAB. The comparison of both the results is shown in this work. First, the open-loop response of the system was simulated to show the behavior of the changes in the input to the output of the engine which is to be controlled. The open-loop response is important in determining the performance of the system and helps in evaluating the performance of the system under different inputs. A comparison of both the controllers is represented which shows that the PID tuned with GA has better performance than the MATLAB proprietary algorithm.

11.5.1 Open-Loop Response of HCCI Engine Model

The open-loop response to the system is shown in Figure 11.2. The figure shows that the system has an undesired response to the input given to it. The response shows the value of CA50 around -0.12 which is not even near to the desired value of the CA50 at the output. The value should be around 5 which is the ideal value of the combustion angle CA50 in the HCCI engines. To compensate for this a controller will be designed which will control the value of the CA50 and bring it back to a value which is stable and is normally the value of all the HCCI engines. The pole zero plot of the system is also shown in Figure 11.3.

The pole zero map is of the NVO as an input and CA50 as an output.

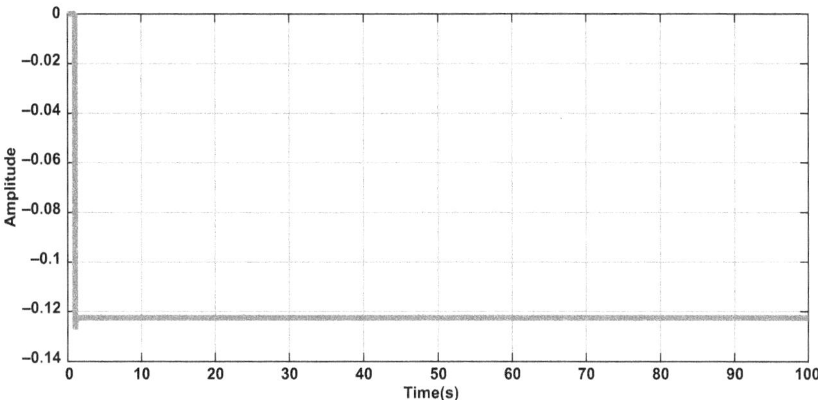

FIGURE 11.2 HCCI engine model open-loop response.

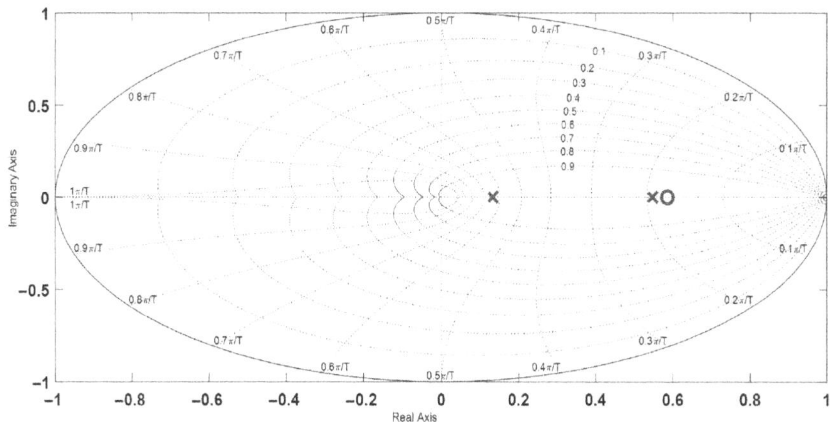

FIGURE 11.3 HCCI engine model pole zero map input NVO and output CA50.

11.5.2 Simulation Results Using MATLAB Proprietary Algorithm

The MATLAB proprietary algorithm was applied to the engine model to tune the PID parameters. The tuning of the results shows that the tuning was done successfully to obtain a stable response from the system, but the system showed very late convergence toward the optimal point. The desired value of 5 CAD was achieved very late by the controller tuned by the proprietary algorithm by MATLAB. The late response shows highly undesirable results since the engine performance is highly dependent upon the on time tracking of the CA50 and any delay in reaching the optimal point for the controller may have a very bad impact on the performance of the engine. The engine works based on the control of combustion phasing and is occurring at proper time. The delay in the combustion phasing has negative effects on the engine performance which may put a stop to the optimal performance of the engine. The MATLAB proprietary algorithm automatically provides the gains of the PID controller using the PID tuner and the user gets a desired response at the output by the working of that controller. The working of the controller is shown in Figure 11.4. The figure shows that the rise time and the settling time of the controller are very high which is highly undesirable since the settling and the rise time need to be less enough so that the engine can perform well under the proper conditions. These parameters make a very important impact on the controller's performance since they will determine the stability and the reference tracking capability of the controller. The PID parameters are tuned and the results are displayed as shown in Figure 11.4.

The input signal is displayed in Figure 11.5. The input is the crank angle at which the exhaust valve closes. This closing of the exhaust valve guarantees the process of negative overlap to happen for the next cycle of the engine. The value of the input signal is around 640–642 and slight changes may occur in this input signal. The reference tracking of the MathWorks proprietary algorithm is shown in Figure 11.6. The figure shows that rise time and the settling time are very high which is not acceptable for the engine operating in recompression engine. The process of NVO may not be fruitful if proper tracking is not ensured.

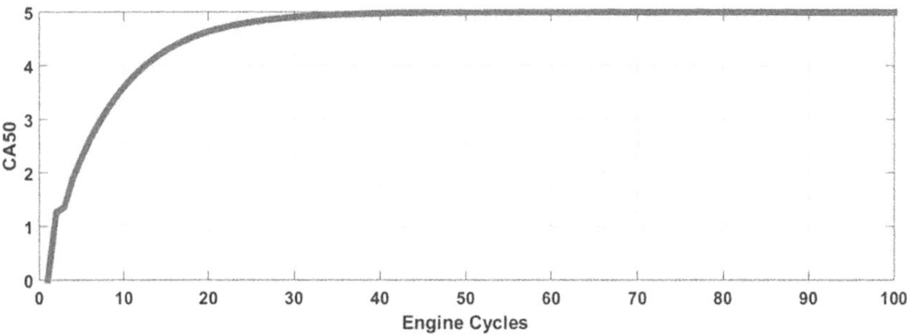

FIGURE 11.4 HCCI engine model closed-loop response using MATLAB proprietary algorithm.

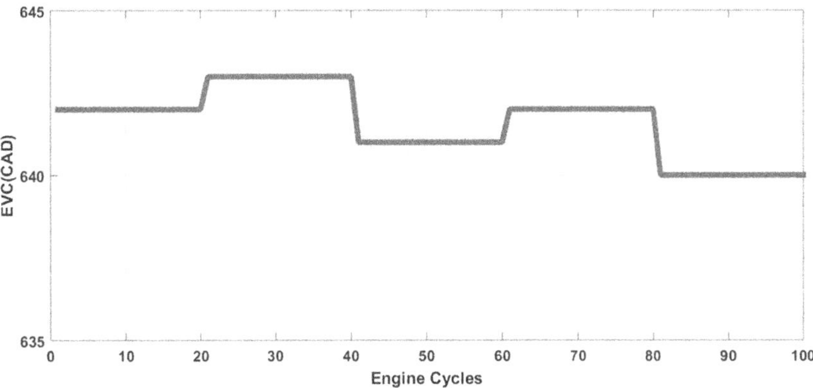

FIGURE 11.5 Input signal.

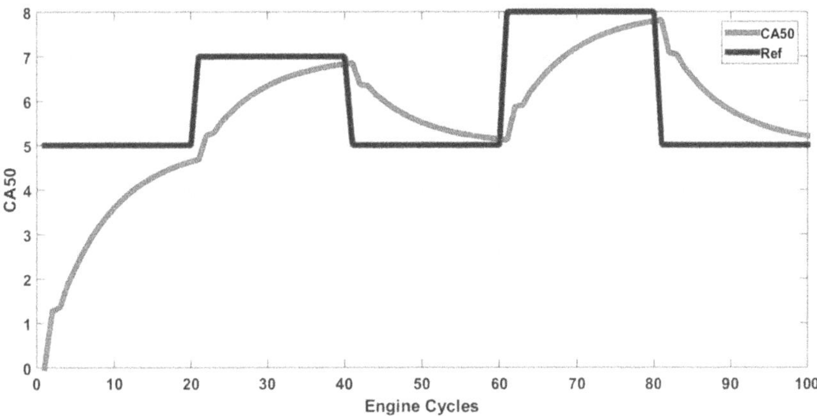

FIGURE 11.6 MATLAB proprietary algorithm reference tracking.

11.5.3 Simulation Results Using PID Controller Tuned with GA

The tuning of PID parameters can be done using several algorithms and techniques. The most widely used methods are the Z-N method and other minimization algorithms which control the performance of the PID parameters and provide a stable output. The usage of different algorithms is recommended in cases where these traditional methods of the tuning do not provide a satisfactory solution for the PID controlled model. The GA provides a very good solution to calculate the desired parameters for the PID controller because the GA-based tuning uses the computation power of the GA to give accurate results at the output. The GA is a very robust technique for the computing of the PID parameters and can cope with different nonlinearities and constrained changes in the plant model which might disturb the actual performance of the HCCI engine. The GA is a better technique for the simulation of the closed-loop engine model because the engine performance is highly dependent upon the performance of the engine. The GA was applied to the model of the HCCI engine with NVO as the input and the CA50 as the output.

The application of the GA was done by implementing the following settings for the iterations of the GA. Figure 11.7 shows the fitness value of the fitness function as the iterations start (Table 11.2). As can be seen clearly from the graphs, the value of the fitness function reduces gradually as the iterations go on. It is an indication of the convergence of the value to the optimum value. The figure clearly shows that the fitness value converges to zero after 5 iterations and the value continues after that. The convergence of the fitness value indicates that the error has been minimized and that the value of error has been reduced which was specified in the fitness function. The error is minimized, and the final goal is achieved for the controller which will tune the PID parameters according to the results of the GA tuning.

Figure 11.8 shows the best values for the three variables used in the optimization process. The process begins with specifying the ranges for the PID parameters and then the algorithm computes the values of the PID parameters based on the convergence of the fitness function to the optimum value. The optimization process

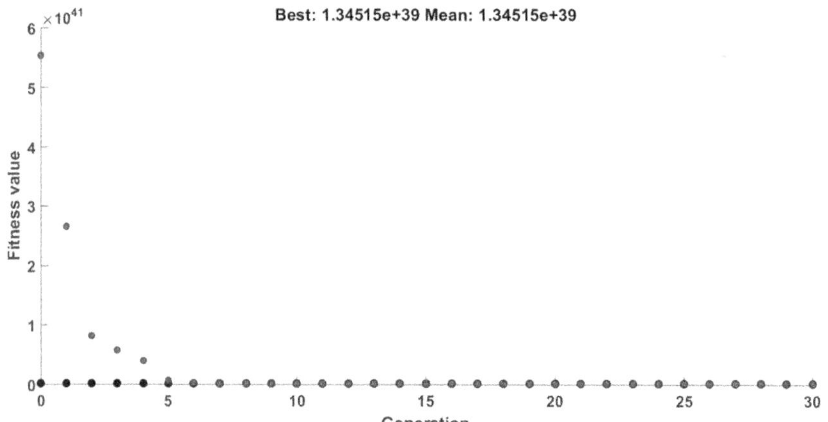

FIGURE 11.7 Best fitness.

TABLE 11.2
GA Parameters Settings

Parameters	Value
Population size	50
Generations	30
Elite count	2.5
Crossover fraction	0.8

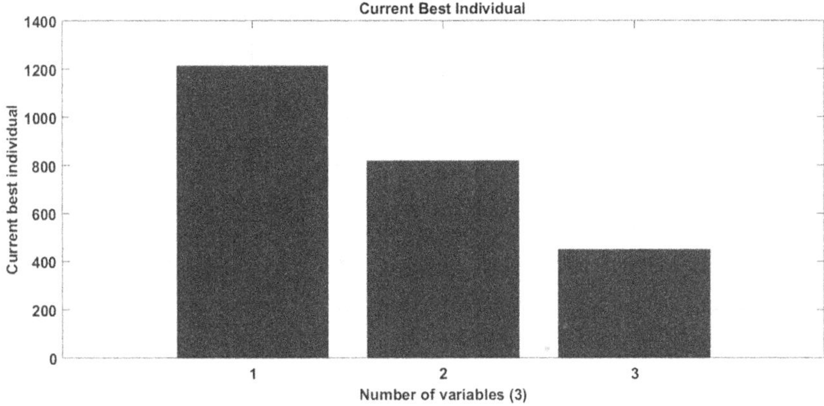

FIGURE 11.8 Best individual.

computes the values of the best PID parameters and is displayed in the form if bar chart. The bar chart explains the best PID values from the entire optimization process. There is constant changing and calculation of these values as the iterations go on and the best values are displayed at the end of the optimization process.

Figure 11.9 shows the performance graph of the GA which shows the average distance between the individuals as the iterations go on. It can be seen clearly that as the iterations increase the distance between the individuals is decreased significantly. It can be seen that after 10 iterations the average distance between the individuals is reduced to zero which means that all of the individuals are converging to a same value and the offsprings in each of the generation are converging to a similar value which will result in the convergence of this optimization problem which is necessary for the solving of the problem.

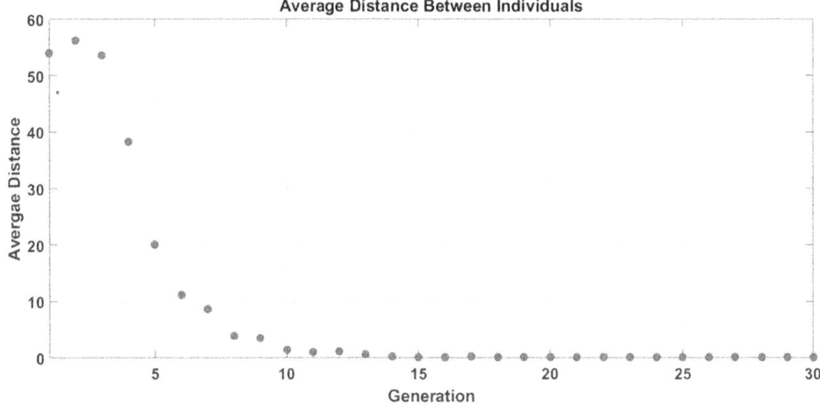

FIGURE 11.9 Distance between individuals.

Figure 11.10 shows the expectation of children over the span of the generations. The figure shows the value of raw scores of the individuals vs the expected number of children. The expected number of children vs their raw scores is shown in this figure.

Figure 11.11 shows the genealogy for the whole optimization process. The genealogy represents all the crossover and mutation processes occurring in the GA over all the generations. The genealogy curve showed most of the blue color which indicates the crossover ratio that was kept in the optimization process. The crossover ratio is kept high which is reflected in the genealogy graph. The dominance of grey color clearly highlights this fact. The red line indicates the mutation ratio in the generations that have occurred over the whole minimization process.

The elite count is usually very low for all the iterations and is kept to a very low value. The elite count for this is very low and is indicated by black lines which occur at the end of each of the generation. After the generations are ended the elite

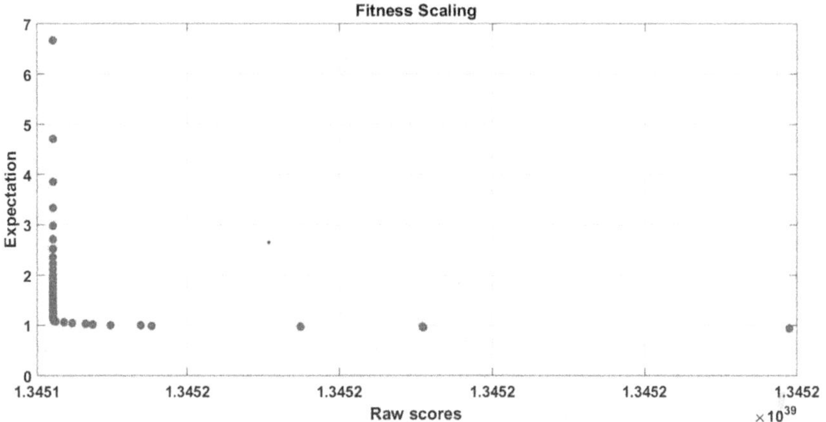

FIGURE 11.10 Expectation.

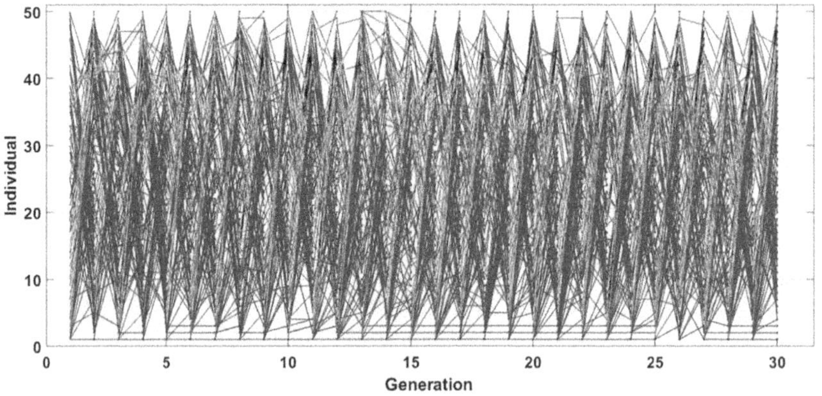

FIGURE 11.11 Genealogy.

individuals are left at the end which are chosen for the next generation which can go on in the next generations and the new generations are created from these newly selected parents.

The best, worst, and the mean scores of the individuals are displayed in Figure 11.12. The figure shows that over the generations the best scores of the individuals occurred at the start of the generations. The generations show that the number of generations increases the mean value eventually converges toward zero, indicating that the value of the individuals is reducing towards zero. The mean value adds up to be zero but the best and the worst scores along with the mean scores are displayed in the figure shown.

The closed-loop response of the engine model is shown in Figure 11.13. The figure shows the closed-loop response for the HCCI engine model. The engine model has the value of CA50 at the output. The value of CA50 as shown has converted to a value of 5 CAD. The results clearly indicate that the value has converged to the optimal value in only a few generations. The convergence shows that the settling time and the rise time for the GA-based PID are better than the proprietary algorithm

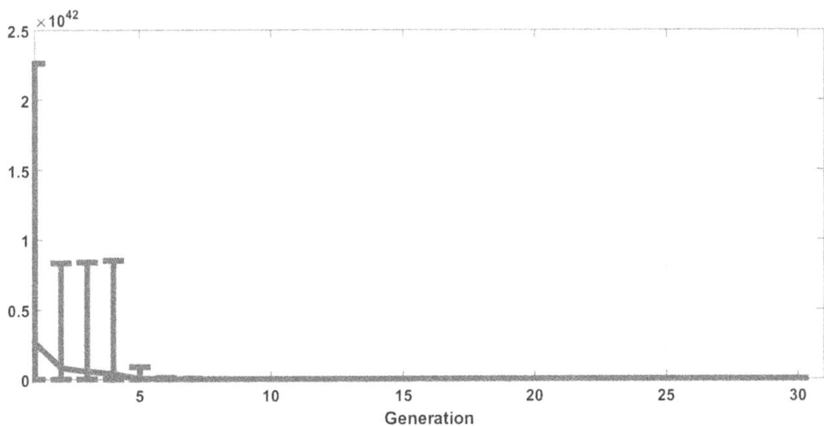

FIGURE 11.12 Best, worst, and mean scores.

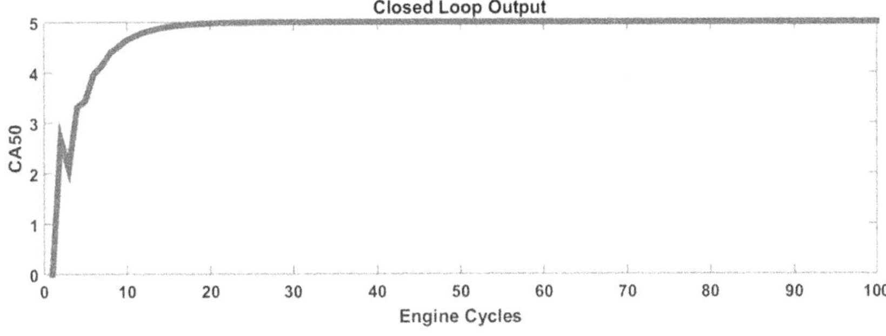

FIGURE 11.13 HCCI engine model closed-loop response using GA-tuned PID controller.

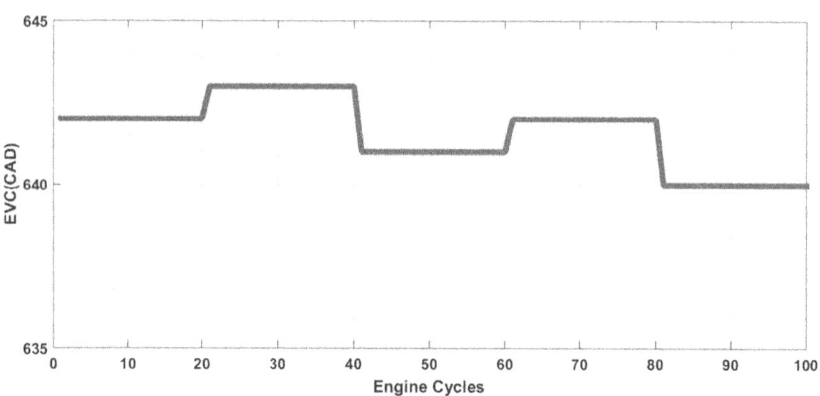

FIGURE 11.14 Input signal.

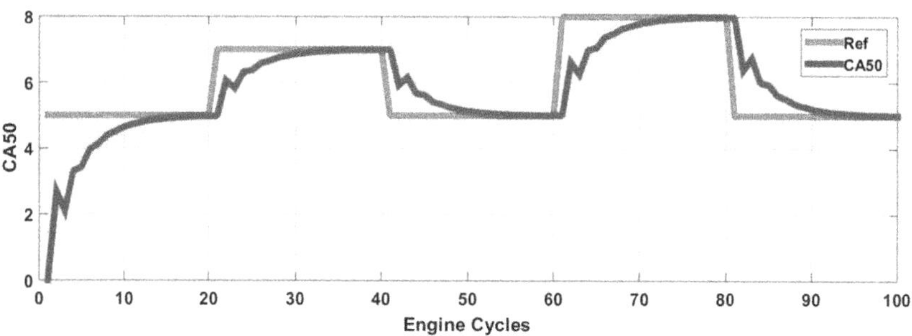

FIGURE 11.15 Reference tracking of the proposed controller.

by MATLAB. These results clearly show the superiority of the GA for the tracking of a particular value. Figure 11.14 shows the change in the reference signal and the robustness of the controller. The graph shows a change in the reference signal over time which is introduced in the system in the form of disturbances. The disturbances change the reference signal, and it helps in the tracking of the signal to the desired value. The results show that the tuning of the PID parameters provided the user with a very good solution in the form of robust controller which can successfully track any changes in the reference signal.

The reference and the controlled output are shown in Figure 11.15. The controller acts immediately and controls the output to a particular value depending on the reference signal that is being fed into the signal.

11.5.4 Performance Comparison of Proposed Controller Vs MathWorks Proprietary Algorithm

The comparison of the proposed controller vs the MathWorks proprietary algorithm is shown in Table 11.3. The table shows that the main difference comes along the rise time and the settling time for both the controllers. The parameters show that the

TABLE 11.3
Tracking Parameters

Specifications	PID(PA-MW)	GA Tuned
Rise time	17 Cycles	6 Cycles
Settling time	30 Cycles	11 Cycles
%Overshoot	0	0

rise time and settling time are very high for the MathWorks Proprietary Algorithm which is not desirable while the GA-tuned PID controller shows better performance in tracking the value of CA50. The rise time and settling time are very much less as compared to the other controller.

11.6 CONCLUSION

NVO technique ensures that the exhaust gases are trapped inside the engine cylinder and because of this the temperature of the intake air is changed as well as the properties of the fuel inside the engine cylinder. Due to these changes, the combustion phasing can be controlled. This chapter described a controller to control the combustion phasing of a recompression HCCI engine. The model was linearized along an operating point to operate a state space model using reverse engineering which was controlled to obtain the desired model which can be controlled. The controller used was a PID controller which was used to control the combustion phasing of the engine. The tuning of the parameters of the engine was done using the GA which is a very good minimization process and can provide the user with excellent results in many cases of controller design. The GA uses the process of natural selection which is very useful for designing robust controllers that can control the output using a very robust approach and can successfully track the reference signal even after variation in it or by introduction of disturbances in the engine. The optimization may be performed using other search algorithms and the results can be compared.

The evaluation of another fitness function which is a close approximation of the original function can also be very interesting. Designing a fitness function based on sinusoidal or Gaussian approximation can be done. This work provides the working of the GA-tuned PID controller and its comparison with the proprietary algorithm from MATLAB. The comparison of results in this chapter shows that as the GA is used for PID tuning it provides better results as compared to the MathWorks proprietary algorithm. The GA considers the optimization technique from the process of natural selection which filters out the bad solutions using the Darwin's theory of natural selection. The theory works on the principle of the survival of the fittest in which the best of the solutions is selected from a current population and are selected for the next generation in which they will reproduce to produce new solutions based on the crossover and the mutation process. The controller can be developed for the nonlinear model of the engine which will help in modelling more details into the model of the engine. The design of nonlinear control helps in determining all the possibilities and the nonlinearities that can occur in the engine model over its operation. The nonlinear controllers consider more complexity into the model which makes the model more robust and realistic as compared to a linearized model which has its own limitations for operating.

REFERENCES

1. Dan Bertolet. Seattles carbon footprint: Assessing the assessment. http://hugeasscity.com/2009/12/13/seattles-carbon-footprint-assessing-the-assessment, 2009.
2. Xavier Tauzia, Alain Maiboom, and Hassan Karaky. Semi-physical models to assess the influence of ci engine calibration parameters on NOx and soot emissions. *Applied Energy*, 208:1505–1518, 2017.
3. Praveen A. Harari, Santosh Ghorpade, Santosh Bhuimbar, Amar S. Kekare, and Amit Deokar. Comprehensive review on emission characteristics of homogeneous charge compression ignition (HCCI) engine operated with renewable fuels. *International Journal of Engineering and Management Research (IJEMR)*, 7(1):375–379, 2017.
4. Tetsuya Kosaka and Toshio Nakanishi. Internal combustion engine, January 13 2015. US Patent D721,097.
5. Shuntaro Okazaki. Internal combustion engine, August 3 2017. US Patent App. 15/329,780.
6. Rakesh Kumar Maurya and Avinash Kumar Agarwal. Combustion and emission characterization of n-butanol fueled HCCI engine. *Journal of Energy Resources Technology*, 137(1):011101, 2015.
7. Can ÇINAR, Ahmet Uyumaz, Seyfi Polat, Emre Yılmaz, Özer Can, and Hamit Solmaz. Combustion and performance characteristics of an HCCI engine utilizing trapped residual gas via reduced valve lift. *Applied Thermal Engineering*, 100: 586–594, 2016.
8. Amin Yousefi, Ayatallah Gharehghani, and Madjid Birouk. Comparison study on combustion characteristics and emissions of a homogeneous charge compression ignition (HCCI) engine with and without pre- combustion chamber. *Energy Conversion and Management*, 100:232–241, 2015.
9. Mustafa Kemal Balki, Cenk Sayin, and Murat Sarıkaya. Optimization of the operating parameters based on taguchi method in an SI engine used pure gasoline, ethanol and methanol. *Fuel*, 180:630–637, 2016.
10. V.K. Shahir, C.P. Jawahar, and P.R. Suresh. Comparative study of diesel and biodiesel on ci engine with emphasis to emissions—A review. *Renewable and Sustainable Energy Reviews*, 45:686–697, 2015.
11. Yulin Chen, Guangyu Dong, J. Hunter Mack, Ryan H. Butt, Jyh-Yuan Chen, and Robert W. Dibble. Cyclic variations and prior-cycle effects of ion current sensing in an HCCI engine: A time-series analysis. *Applied Energy*, 168:628–635, 2016.
12. Willard W. Pulkrabek. *Engineering Fundamentals of the Internal Combustion Engine*. Pearson Prentice Hall, Upper Saddle River, NJ, 2014.
13. Can Cinar, Ahmet Uyumaz, Hamit Solmaz, and Tolga Topgul. Effects of valve lift on the combustion and emissions of a HCCI gasoline engine. *Energy Conversion and Management*, 94:159–168, 2015.
14. Chunsheng Ji, John E. Dec, Jeremie Dernotte, and William Cannella. Effect of ignition improvers on the combustion performance of regular-grade e10 gasoline in an HCCI engine. *SAE International Journal of Engines*, 7(2014-01-1282):790–806, 2014.

15. Samveg Saxena, Nihar Shah, Ivan Bedoya, and Amol Phadke. Understanding optimal engine operating strategies for gasoline-fueled HCCI engines using crank-angle resolved exergy analysis. *Applied Energy*, 114:155–163, 2014.
16. Bang-Quan He, Mao-Bin Liu, and Hua Zhao. Comparison of combustion characteristics of n-butanol/ethanol–gasoline blends in a HCCI engine. *Energy Conversion and Management*, 95:101–109, 2015.
17. A. Selvakumar, M. Senthil, V. Prabhakaran, S. Pavithran, S. Keshav, et al. Development and testing of variable valve actuation (VVA) system in homogeneous charge compression ignition (HCCI) engine. *Advances in Natural and Applied Sciences*, 10(9 SE):-11–19, 2016.
18. Mao-Bin Liu, Bang-Quan He, and Hua Zhao. Effect of air dilution and effective compression ratio on the combustion characteristics of a HCCI (homogeneous charge compression ignition) engine fuelled with n-butanol. *Energy*, 85:296–303, 2015.
19. Seyfi Polat. An experimental study on combustion, engine performance and exhaust emissions in a HCCI engine fuelled with diethyl ether–ethanol fuel blends. *Fuel Processing Technology*, 143:140–150, 2016.
20. Jan-Ola Olsson, Per Tunestål, and Bengt Johansson. Closed-loop control of an HCCI engine. Technical report, SAE Technical Paper, 2001.
21. Fredrik Agrell, Hans-Erik Ångström, Bengt Eriksson, Jan Wikander, and Johan Linderyd. Integrated simulation and engine test of closed loop HCCI control by aid of variable valve timings. Technical report, SAE Technical Paper, 2003.
22. Jason S. Souder, J. Hunter Mack, J. Karl Hedrick, and Robert W. Dibble. Microphones and knock sensors for feedback control of HCCI engines. Paper No. ICEP2004-960, 2004.
23. Fredrik Agrell, Hans-Erik Ångström, Bengt Eriksson, Jan Wikander, and Johan Linderyd. Transient control of HCCI through combined intake and exhaust valve actuation. Technical report, SAE Technical Paper, 2003.
24. Göran Haraldsson, Per Tunestål, Bengt Johansson, and Jari Hyvönen. Transient control of a multi cylinder HCCI engine during a drive cycle. Technical report, SAE Technical Paper, 2005.
25. Shaver GM, and Gerdes, JC. Cycle-to-Cycle Control of HCCI Engines. *Proceedings of the ASME 2003 International Mechanical Engineering Congress and Exposition.* Dynamic Systems and Control, Vols. 1 and 2. Washington, DC, USA. November 15–21, pp. 403–412, 2003. ASME. https://doi.org/10.1115/IMECE2003-41966.
26. Chia-Jui Chiang and A.G. Stefanopoulou. Control of thermal ignition in gasoline engines. In *American Control Conference, 2005. Proceedings of the 2005*, pp. 3847–3852. IEEE, 2005.
27. Nikhil Ravi, Hsien-Hsin Liao, Adam F. Jungkunz, Anders Widd, and J. Christian Gerdes. Model predictive control of HCCI using variable valve actuation and fuel injection. *Control Engineering Practice*, 20(4):421–430, 2012.
28. Chia-Jui Chiang and Chian-Ling Chen. Constrained control of homogeneous charge compression ignition (HCCI) engines. In *Industrial Electronics and Applications (ICIEA), 2010 the 5th IEEE Conference on*, pp. 2181–2186. IEEE, 2010.
29. Johan Bengtsson, Petter Strandh, Rolf Johansson, Per Tunestål, and Bengt Johansson. Hybrid modelling of homogeneous charge compression ignition (HCCI) engine dynamicsa survey. *International Journal of Control*, 80(11):1814–1847, 2007.
30. Nikhil Ravi, Hsien-Hsin Liao, Adam F. Jungkunz, Chen-Fang Chang, Han Ho Song, and J. Christian Gerdes. Modeling and control of an exhaust recompression HCCI engine using split injection. *Journal of Dynamic Systems, Measurement, and Control*, 134(1):011016, 2012.

31. Maria Karlsson, Kent Ekholm, Petter Strandh, Rolf Johansson, Per Tunestål, and Bengt Johansson. Closed-loop control of combustion phasing in an HCCI engine using VVA and variable EGR. *IFAC Proceedings Volumes*, 40(10):501–508, 2007.
32. Gregory M. Shaver, J. Christian Gerdes, and Matthew Roelle. Physics-based closed-loop control of phasing, peak pressure and work output in HCCI engines utilizing variable valve actuation. In *American Control Conference, 2004. Proceedings of the 2004*, vol.1, pp. 150–155. IEEE, 2004.
33. Nick J. Killingsworth, Salvador M. Aceves, Daniel L. Flowers, and Miroslav Krstic. Extremum seeking tuning of an experimental HCCI engine combustion timing controller. In *American Control Conference, 2007. ACC'07*, pp. 3665–3670. IEEE, 2007.
34. D.J. Rausen, A.G. Stefanopoulou, J.-M. Kang, J.A. Eng, and T.-W. Kuo. A mean-value model for control of homogeneous charge compression ignition (HCCI) engines. *Journal of Dynamic Systems, Measurement, and Control*, 127(3):355–362, 2005.
35. Christopher G. Mayhew, Karl Lukas Knierim, Nalin A. Chaturvedi, Sung-bae Park, Jasim Ahmed, and Aleksandar Kojic. Reduced-order modeling for studying and controlling misfire in four-stroke HCCI engines. In *Decision and Control, 2009 Held Jointly with the 2009 28th Chinese Control Conference. CDC/CCC 2009. Proceedings of the 48th IEEE Conference on*, pp. 5194–5199. IEEE, 2009.
36. Daniel Blom, Maria Karlsson, Kent Ekholm, Per Tunestål, and Rolf Johansson. HCCI engine modeling and control using conservation principles. Technical report, SAE Technical Paper, 2008.
37. Aristotelis Babajimopoulos, V.S.S.P. Challa, George A. Lavoie, and Dennis N. Assanis. Model-based assessment of two variable cam timing strategies for HCCI engines: Recompression vs. rebreathing. *Ann Arbor*, 1001:48109, 2009.
38. Robert Szolak, Eric Alexander Morales Wiemer, Ivica Kraljevic, Alexander Susdorf, Hüseyin Karadeniz, Boris Epple, Florian Rümmele, and Achim Schaadt. On-board fuel tailoring with a novel catalytic evaporator for HCCI combustion. In *SAE Technical Paper*. SAE International, 04 2016.
39. Karthikayan Sundararajan, Krishnaraj Janathanan, Vasanthakumar Pandian, Madhankumar Dhandapani, and Kalaiyarasan Kanagara. A performance, combustion and emission study on HCCI engine: Trends and innovations. In *SAE Technical Paper*. SAE International, 02 2016.
40. Akhilendra Pratap Singh and Avinash Kumar Agarwal. Effect of intake charge temperature and EGR on biodiesel fuelled HCCI engine. In *SAE Technical Paper*. SAE International, 02 2016.
41. Jean-Baptiste Masurier, Fabrice Foucher, Guillaume Dayma, Christine Rousselle, and Philippe Dagaut. Application of an ozone generator to control the homogeneous charge compression ignition combustion process. In *SAE Technical Paper*. SAE International, 09 2015.
42. Pietro Matteo Pinazzi, Jean-Baptiste Masurier, Guillaume Dayma, Philippe Dagaut, and Fabrice Foucher. Towards stoichiometric combustion in HCCI engines: Effect of ozone seeding and dilution. In *SAE Technical Paper*. SAE International, 09 2015.
43. Mark Aaron Hoffman and Zoran Filipi. Influence of directly injected gasoline and porosity fraction on the thermal properties of HCCI combustion chamber deposits. In *SAE Technical Paper*. SAE International, 09 2015.
44. Tatsuya Kuboyama, Shunsuke Goto, Yasuo Moriyoshi, Keiichi Koseki, and Yoichi Akiyama. Effect of low octane gasoline on performance of a HCCI engine with the blowdown supercharging. In *SAE Technical Paper*. SAE International, 09 2015.
45. Hiroki Takeori, Hiroki Hosoe, Teruyoshi Morita, and Tetsuo Endo. A study of aftertreatment system for spark-assisted HCCI engine. In *SAE Technical Paper*. SAE International, 09 2015.

46. Alvaro Pinheiro, David Vuilleumier, Darko Kozarac, and Samveg Saxena. Simulating a complete performance map of an ethanol-fueled boosted HCCI engine. In *SAE Technical Paper*. SAE International, 04 2015.
47. Göran Haraldsson, Per Tunestål, Bengt Johansson, and Jari Hyvönen. HCCI closed-loop combustion control using fast thermal management. Technical report, SAE Technical Paper, 2004.
48. Jari Hyvönen, Göran Haraldsson, Bengt Johansson. Balancing cylinder-to-cylinder variations in a multi-cylinder VCR-HCCI engine. Technical report, SAE Technical Paper, 2004.
49. Joel Martinez-Frias, Salvador M. Aceves, Daniel Flowers, J. Ray Smith, and Robert Dibble. HCCI engine control by thermal management. Technical report, SAE Technical Paper, 2000.
50. Jialin Yang, Todd Culp, and Thomas Kenney. Development of a gasoline engine system using HCCI technology-the concept and the test results. Technical report, SAE Technical Paper, 2002.
51. Fuquan Zhao, Thomas N. Asmus, Dennis N. Assanis, John E. Dec, James A. Eng, and Paul M. Najt. Homogeneous charge compression ignition (HCCI) engines. Technical report, SAE Technical Paper, 2003.
52. Rakesh Kumar Maurya and Avinash Kumar Agarwal. Experimental investigation on the effect of intake air temperature and air–fuel ratio on cycle-to-cycle variations of HCCI combustion and performance parameters. *Applied Energy*, 88(4):1153–1163, 2011.
53. M. Lida, M. Hayashi, D.E. Foster, and J.K. Martin. Characteristics of homogeneous charge compression ignition (HCCI) engine operation for variations in compression ratio, speed, and intake temperature while using n-butane as a fuel. *Journal of Engineering for Gas Turbines and Power*, 125(2):472–478, 2003.
54. Göran Haraldsson, Per Tunestål, Bengt Johansson, and Jari Hyvönen. HCCI combustion phasing in a multi cylinder engine using variable compression ratio. Technical report, SAE Technical Paper, 2002.
55. Göran Haraldsson, Per Tunestål, Bengt Johansson, and Jari Hyvönen. HCCI combustion phasing with closed-loop combustion control using variable compression ratio in a multi cylinder engine. Technical report, SAE Technical Paper, 2003.
56. Koji Hiraya, Kazuya Hasegawa, Tomonori Urushihara, Akihiro Iiyama, and Teruyuki Itoh. A study on gasoline fueled compression ignition engine a trial of operation region expansion. Technical report, SAE Technical Paper, 2002.
57. Hatim Machrafi, Simeon Cavadias, and Philippe Guibert. An experimental and numerical investigation on the influence of external gas recirculation on the HCCI autoignition process in an engine: Thermal, diluting, and chemical effects. *Combustion and Flame*, 155(3):476–489, 2008.
58. Magnus Sjöberg, John E. Dec, and Wontae Hwang. Thermodynamic and chemical effects of EGR and its constituents on HCCI autoignition. Technical report, SAE Technical Paper, 2007.
59. Tian Guohong, Wang Zhi, Wang Jianxin, Shuai Shijin, and An Xin-Liang. HCCI combustion control by injection strategy with negative valve overlap in a GDI engine. Technical report, SAE Technical Paper, 2006.
60. Patrick A. Caton, Aaron J. Simon, J. Christian Gerdes, and Christopher F. Edwards. Residual-effected homogeneous charge compression ignition at a low compression ratio using exhaust reinduction. *International Journal of Engine Research*, 4(3):163–177, 2003.
61. Thomas Johansson, Bengt Johansson, Per Tunestål, and Hans Aulin. HCCI operating range in a turbo-charged multi cylinder engine with VVT and spray-guided DI. Technical report, SAE Technical Paper, 2009.

62. Jacob R. Zuehl, Jaal Ghandhi, Christopher Hagen, and William Cannella. Fuel effects on hcci combustion using negative valve overlap. In *SAE Technical Paper.* SAE International, 04 2010.
63. Martin Larsson, Ingemar Denbratt, and Lucien Koopmans. Ion current sensing in an optical HCCI engine with negative valve overlap. In *SAE Technical Paper.* SAE International, 01 2007.
64. Lucien Koopmans, Roy Ogink, and Ingemar Denbratt. Direct gasoline injection in the negative valve overlap of a homogeneous charge compression ignition engine. In *SAE Technical Paper.* SAE International, 05 2003.
65. Patrick Borgqvist, Martin Tuner, Augusto Mello, Per Tunestål, and Bengt Johansson. The usefulness of negative valve overlap for gasoline partially premixed combustion, PPC. In *SAE Technical Paper.* SAE International, 09 2012.
66. Takeru Ibara, Minoru Iida, and David E. Foster. Study on characteristics of gasoline fueled HCCI using negative valve overlap. In *SAE Technical Paper.* SAE International, 11 2006.
67. Russell P. Fitzgerald, Richard Steeper, Jordan Snyder, Ronald Hanson, and Randy Hessel. Determination of cycle temperatures and residual gas fraction for HCCI negative valve overlap operation. *SAE International Journal of Engines*, 3:124–141, 04 2010.
68. Tobias Joelsson, Rixin Yu, Johan Sjholm, Per Tunestål, and Xue-Song Bai. Effects of negative valve overlap on the auto-ignition process of lean ethanol/air mixture in HCCI-engines. In *SAE Technical Paper.* SAE International, 10 2010.
69. Nikhil Ravi, Hsien-Hsin Liao, Adam F. Jungkunz, and J. Christian Gerdes. Modeling and control of exhaust recompression HCCI using split injection. In *American Control Conference (ACC), 2010*, pp. 3797–3802. IEEE, 2010.
70. Ribeiro JMS, Santos MF, Carmo MJ and Silva MF, Comparison of PID controller tuning methods: analytical/classical techniques versus optimization algorithms, In *2017 18th International Carpathian Control Conference (ICCC)*, Sinaia, pp. 533–538, 2017, doi: 10.1109/CarpathianCC.2017.7970458
71. Karl J. Åström and Tore Hägglund. *Advanced PID Control.* ISA-The Instrumentation, Systems and Automation Society, 2006.
72. Sigurd Skogestad. Simple analytic rules for model reduction and PID controller tuning. *Journal of Process Control*, 13(4):291–309, 2003.
73. Zwe-Lee Gaing. A particle swarm optimization approach for optimum design of PID controller in AVR system. *IEEE Transactions on Energy Conversion*, 19(2):384–391, 2004.
74. Ibrahim, Muhammed A., Ausama Kh Mahmood, Nashwan Saleh Sultan. Optimal PID controller of a brushless DC motor using genetic algorithm. *International Journal of Power Electronics and Drive Systems*, ISSN 2088.8694, 8694, 2019.
75. Jinhua Zhang, Jian Zhuang, Haifeng Du, and Wang Sun'an. Self-organizing genetic algorithm-based tuning of PID controllers. *Information Sciences*, 179(7):1007–1018, 2009.
76. Dionisio S. Pereira and João O.P. Pinto. Genetic algorithm-based system identification and pid tuning for optimum adaptive control. In *Advanced Intelligent Mechatronics. Proceedings, 2005 IEEE/ASME International Conference on*, pp. 801–806. IEEE, 2005.
77. Q. Cao and L. Chang, Genetic algorithm optimization for high-performance VSI-Fed permanent magnet synchronous motor drives. In *2006 37th IEEE Power Electronics Specialists Conference*, Jeju, Korea (South), pp. 1–7, 2006, doi: 10.1109/pesc.2006.1711959.

78. Tan Chaodong, Tan Pengfei, Li Xinlun, Wu Haoda, and Yang Ruogu. Research on flexible variable-speed control model and optimization method of rod pumping well based on genetic algorithm. In *Applied System Innovation (ICASI), 2017 International Conference on*, pp. 1771–1774. IEEE, 2017.
79. P. Kongsereeparp and M. David Checkel. Study of reformer gas effects on n-heptane HCCI combustion using a chemical kinetic mechanism optimized by genetic algorithm. Technical report, SAE Technical Paper, 2008.
80. Jing Yang, Zhixiong Zhang, Yi Wang, et al. Optimization of a high-speed gasoline engine using genetic algorithm. Technical report, SAE Technical Paper, 2013.
81. Leibnitz Germanio and Ricardo Luiz Utsch de Freitas Pinto. Optimization of a cam by a genetic algorithm. Technical report, SAE Technical Paper, 2002.
82. Mahdi Ahmadi. Intake, exhaust and valve timing design using single and multi-objective genetic algorithm. Technical report, SAE Technical Paper, 2007.
83. Longchen Li, Wei Huang, Hailin Ruan, Xiujie Tian, Keda Zhu, Melvyn Care, Richard Wentzel, Xiaojun Chen, and Changwei Zheng. A new strategy optimization method for vehicle active noise control based on the genetic algorithm. Technical report, SAE Technical Paper, 2017.
84. Yanzhe Sun, Tianyou Wang, Zhen Lu, Lei Cui, and Ming Jia. The optimization of intake port using genetic algorithm and artificial neural network for gasoline engines. Technical report, SAE Technical Paper, 2015.
85. Hai-Wen Ge, Yu Shi, Rolf D. Reitz, David D. Wickman, Guangsheng Zhu, Houshun Zhang, and Yury Kalish. Heavy-duty diesel combustion optimization using multi-objective genetic algorithm and multi-dimensional modeling. In *SAE Technical Paper*. SAE International, 04 2009.
86. M.R. Bolhasani and Sh. Azadi. Parameter estimation of vehicle handling model using genetic algorithm. In *SAE Technical Paper*. SAE International, 05 2002.
87. Peng Liu, Liyun Fan, De Xu, Xiuzhen Ma, and Enzhe Song. Multi-objective optimization of high-speed solenoid valve based on response surface and genetic algorithm. In *SAE Technical Paper*. SAE International, 04 2015.

12 Fabrication of Smart-Meter for Power Consumption Measurements of Machine Tools

Aqib Mashood khan, Ning He, Wei Zhao, Cheng Zhang, and Muhammad Jamil
Nanjing University of Aeronautics and Astronautics

CONTENTS

12.1 INTRODUCTION

In recent years, with the increasing severity of energy consumption and environmental problems in manufacturing, the call for sustainable manufacturing is getting higher and higher, and traditional processing and manufacturing industries will inevitably carry out industrial upgrading and technological innovation with the goals of green, clean, and energy saving. Globally, manufacturing energy consumption accounts for 37% of global basic energy consumption [1], and energy demand is increasing. Energy is an important guarantee for future economic growth and prosperity, and an important indicator of global competitiveness. Therefore, improving the energy efficiency of the manufacturing industry has become one of the global strategic goals. The main challenges faced are the rapid growth of energy costs and how to reduce carbon emissions [2]. The BP World Energy Statistical Yearbook 2018 analyzed global energy consumption. In 2017, China accounted for 23.2% of global energy consumption, a year-on-year increase of 3.1%, and the growth of CO_2 emissions of 1.6% [3]. Continuously increasing energy consumption has provided an important guarantee for the rapid growth of China's economy, but the current industrial development still relies on the extensive model of high energy consumption and high carbon emissions. Therefore, saving energy and improving energy efficiency are of great significance to the sustainable development of our economy.

During industrial development, China has always attached great importance to resource conservation and ecological environmental protection and adhered to the basic national policy of resource conservation and environmental protection. The report of the 15th National Congress of the Communist Party of China clearly put forward the implementation of sustainable development strategy. The 17th National Congress of the CPC emphasized that by 2020, an industrial structure, growth mode, and consumption model that will save energy resources and protect the ecological environment will basically be formed [4]. The report of the 18th National Congress of the Communist Party of China promotes sustainable development to the height of green development, further promotes industrial transformation and upgrading, and actively leads emerging industries to a high starting point for green development. In March 2016, the "Outline of the Thirteenth Five-Year Plan for National Economic and Social Development of the People's Republic of China" was issued. The document states that it is necessary to comprehensively promote energy conservation, effectively control greenhouse gas emissions, and establish and improve the efficient use of resources. The determination to accelerate the upgrading of the low-end manufacturing industry responds to global climate change and controls the greenhouse effect. In this context, a comprehensive and systematic analysis of manufacturing energy consumption is proposed; proposing effective energy-saving methods is a key step in reducing greenhouse gases [5].

The machine tool (MT) is a platform for industrial production, and it is also the main body of energy consumption during processing. There are many sources of energy consumption for MTs, and the changes in energy consumption are complicated. Worldwide, the measurement of MT energy consumption and the study of energy characteristics have received widespread attention from scholars. In developed countries and regions such as Europe, the United States, and Japan, the successive introduction of energy consumption directives and the introduction of carbon taxes have greatly promoted the related research on MT energy consumption. China is the world's largest MT production and consumption market [6], and the number of MTs is as high as 8 million [7]. However, MT equipment technology still has a certain gap compared with developed countries. Energy efficiency is not high, and the impact on the environmental issues such as waste of resources has not received enough attention. Therefore, MTs produced in China often fail to meet the green and energy-saving targets of developed countries, limiting the market for domestic MTs. Process parameters such as depth of cut, cutting speed, feed rate, and width of cut are commonly varied to study the variation in cutting power and energy consumption of MT.

To sum up, research on the energy consumption of MTs is of great significance for achieving green manufacturing. It needs to work in the following three aspects: (i) characterize and model the energy consumption of the machining process; (ii) obtain the actual machining energy of the MT consumption and carbon emissions information; (iii) reduce MT energy consumption and reduce carbon emissions and environmental impact. MT energy consumption modeling is recognized as an effective method and evaluation standard for MT energy consumption [8], and it is an important basis for developing energy-saving potentials. Based on the establishment of the MT energy consumption model, designing and developing an energy consumption monitoring system and obtaining dynamic information about the actual processing energy sources are the prerequisites for achieving optimization of processing parameters, energy-saving, and emission reduction. At present, a commonly used method for monitoring MT energy consumption is to use a power meter or an information acquisition card to connect the MT for energy consumption measurement, but the scalability is poor and the cost is high. The other is to directly read the energy consumption data of a CNC MT or scan the CNC code to calculate the energy consumption information through the software. This software-based energy consumption monitoring method is complicated in operation and poor in versatility. In view of the above problems, this chapter designs and develops a MT energy consumption monitoring system with high-cost performance, a high degree of interaction, and good scalability.

12.2 LITERATURE REVIEW

There are many energy sources and complex structures in the machining process of MTs. The energy consumption analysis involves the study of MT structure and tool parameters in the field of mechanical processing, and the study of motor parameters and control systems in the field of power electronics. Therefore, the modeling of MT energy consumption is a comprehensive interdisciplinary problem, and domestic and foreign scholars have carried out a lot of research work from multiple levels of MT energy consumption.

12.2.1 Research on Cutting Energy Consumption Model

The earliest analysis of MT energy consumption was carried out from the perspective of cutting force. In 1945, American Scholar Merchant established the cutting force formula, which described the relationship between cutting force and cutting parameters, tool and workpiece material properties, and the cutting power. Discussion and analysis were also carried out, and a method for calculating cutting energy consumption by cutting force and cutting speed was proposed [9]. In 2000, H. M. Ertunc et. al. proposed a functional relationship between tool wear and cutting energy consumption, and provided a procedure for measuring cutting power [10]. Hossein researched the cutting force during the milling process of ISI P20 tool steel and established the relationship between the cutting force and the cutting parameters through statistical analysis [11]. In addition, some scholars have studied the energy consumption model of MTs from the perspective of thermodynamics. Dahmus has established a thermodynamic energy model of machining processes. Studies show that the actual cutting energy consumption only accounts for about 15% of the total machining energy consumption, and the brake is activated. The energy consumption of stages such as air travel and empty stroke accounts for the main part of MT energy consumption [12]. Li Congbo studied the hobbing machining technology, analyzed the energy consumption characteristics of the machining process, and established a comprehensive energy consumption model with the spindle speed and feed force as key variables. Li proposed an improved energy consumption model based on the heat balance equation and empirical formula. This model describes the relationship between material removal rate, spindle speed, and energy consumption. It is verified by milling experiments that the model can control the cutting process through the process parameters of the machining process and perform reliable energy consumption predictions [13].

12.2.2 Research on Energy Consumption Model of MT Component Layer

MT energy consumption components involve multiple complex systems. When analyzing MT energy consumption, each energy consumption component needs to be studied separately. From the perspective of the integration of the spindle motor and the mechanical transmission system, Professor Liu Fei of Chongqing University in China conducted a decomposition study of the machining system for the first time. The analysis was mainly from the two aspects of the main drive system and the feed drive system of the MT, and the energy flow of the MT was proposed. The concept [14] analyzes and expresses the energy loss and kinetic energy conversion of the MT's main drive system, and establishes a mathematical model of the energy transmission of the main machine's main drive system.

12.2.3 Developed and Used Instrument: A Smart Meter

In recent days, awareness about energy consumption in the metal processing industry is increasing. However, still, modeling and characterization of energy consumption of MTs during machining of difficult to cut machining have not been studied.

It is impossible to reduce/save energy consumption without assessing energy consumption in the mechanical machining process. There are various tools and methods available to assess the energy consumption of MTs during the machining process. Commercially available real-time power monitoring systems to measure power consumption are expensive. Thus, a low-cost power measurement system is necessary to be developed.

A new custom-built low-cost energy measurement hardware called *smart meter* was used to measure the power consumption during different stages of the MT. The smart meter basically consists of Raspberry Pi-3, Smart Pi Expansion Module (SPEM), Keyboard and Mouse, HDMI connector, Micro SD Card. Raspberry Pi-3 is a potent processor and it possesses features such as Quad-Core 1.2 GHz Broadcom 64 bit CPU, 1 GB RAM, wireless LAN, and Bluetooth low energy. The most important function lies in extended 40-pin GPIO header, which can be used for serial buses like SPI, I2C, 1-Wire, etc. Currents can be measured up to 300 A via inductive current sensors in the custom-built smart meter. SmartPi module was designed to have screw terminals to measure voltage in the range of 400 V on all phases. Hereby, for example, the direction of energy flow is determinable. First, idle power or standby power was measured when the MT was turned on but spindle and a table was at rest. Second, cutting power was measured at different spindle speeds. Third, air cutting power was measured while certain spindle speed and table movement, but without cutting the workpiece. Lastly, machining power was measured while all movements of machine accessories while cutting the workpiece.

Smart meter accessed the acquired power data by a new approach using low-threshold programming software, which enables to create its customized user interface (UI). Node-RED is a flow-based programming tool for the Internet of Things (IoT) and it provides Brower-based flow editor to record three-phase current signals and store measured data in Raspberry Pi as CSV file. Moreover, data will upload with the MQTT transmission protocol for remote access. It is worth mentioning that the measurement accuracy of the smart meter was validated by comparing its data with commercially used Load Controls PPC-3 Power Meter that was readily available [15].

12.3 OVERALL SCHEME DESIGN OF THE MONITORING SYSTEM

Based on the analysis of different processing states and energy-consuming components of MTs, the main objectives were to develop a low-cost energy measurement setup that could measure the load energy consumption of MTs and the energy consumption of the main drive system of MTs. The hardware of the MT energy consumption monitoring system needs to realize two main functions: (i) signal acquisition and real-time transmission of sensors; (ii) data display and storage. The hardware of the system was based on open-close current mutual inductance technology, integrated circuit technology of electric energy measurement, and ARM core microcomputer technology. The monitoring requirement was realized through the cooperation of three-hardware modules: Sensor, *Raspberry Pi*, and SPEM development board.

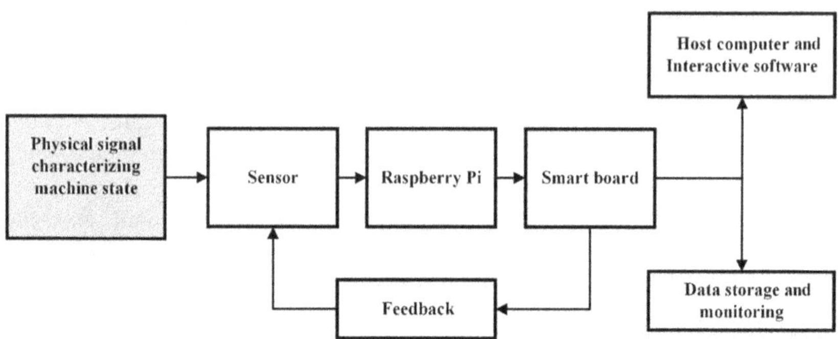

FIGURE 12.1 Schematic diagram of the working principle of smart meter.

12.4 WORKING PRINCIPLE DESIGN OF MONITORING SYSTEM

The working principle of the monitoring system was that the real-time power signal of MT processing was acquired by sensors, and the signal amplification, analog-to-digital conversion, and digital signal processing were completed by the IC technology of energy measurement integrated by the lower computer.

The processed digital signal was transmitted from the SPEM to the *Raspberry Pi* for further processing. Finally, *Raspberry Pi* displayed and stored the data. The signal transmission process of this system is shown in Figure 12.1. The Smart Pi expansion-module (SPEM) acquisition unit received the signal and processed it. If the error was large, it sent it back to the sensor to adjust the power values. The feedback system helps the smart sensor to reduce the error in final power values.

12.5 DESIGN OBJECTIVES AND REQUIREMENTS OF THE MONITORING SYSTEM

According to the functions of the monitoring system, the technical design indexes of the system are as follows:

(1) The current transformer needs to realize the acquisition function of the current signal. It should be able to receive a three-phase current signal at the same time. The range should be greater than 10 A, and the sampling frequency should be higher than 50 Hz. (2) The electric energy meter should be compatible with three-phase, three-wire, or four-wire configuration, and has the function of analog-to-digital conversion. High-speed data acquisition port is equipped to realize digital signal output and real-time power information reading. (3) *Raspberry Pi* has the function of reading the electric energy meter signal through the communication port to realize data processing and storage, and has certain computing power and expansibility to display the processed data dynamically and realize the function of uploading data. The stable power consumption of each electronic component needs to be maintained at a low level. (4) On the basis of realizing the monitoring function, the performance-price ratio of hardware and software of the system should be higher, and the cost should be lower than the current general energy consumption monitoring system scheme.

According to the current research status and development trend, the monitoring system meets the following requirements in design:

12.5.1 Advancement

In the industrial application of MTs, the commonly used hardware platforms are power meters of the various manufacturers and data acquisition cards are used for recording energy consumption information at high frequency. These two hardware platforms have their own advantages and disadvantages in practical application. Power meters can achieve simple data acquisition, but they need supporting software systems. Their scalability is significantly limited, and user interaction is not strong. The system based on the data acquisition card directly connected to the computer to develop a monitoring system is complex. It requires strong professional knowledge to analyze the sampling results, and the cost of hardware and software is high.

To solve the above problems, *Raspberry Pi* is a microcomputer based on the ARM core. The price is only 350 yuan. The price of the development board of the SPEM is 1200 yuan. The overall cost of the system is less than 2000 yuan. The price of industrial data acquisition card is as high as tens of thousands, and the overall cost of the system is higher. Compared with the common development board, *Raspberry Pi* has more powerful computing power and good expansibility, to achieve stable data acquisition and storage. At the same time, it is equipped with high-speed data acquisition ports and ports supporting serial transmission protocols to facilitate data upload. After the connection between the *Raspberry Pi* and the SPEM was completed, it was encapsulated in a plastic shell. At the same time, the two main boards were fixed with double-headed bolts and nuts to avoid short circuits when they are in direct contact. The encapsulated monitoring system can be fixed and disassembled conveniently. The power meter used in the industry is often bulky and has poor portability. When the system is in the network environment, the design of the remote-control module can make the operation of the monitoring system more convenient.

12.5.2 Scalability

In the proposed system, the power information of the main energy-consuming parts of the MT was acquired in real-time by the noncontact current transformer, and other analog or digital signals such as temperature and humidity can also be monitored by various ports reserved by the development board. Because the ARM core microcomputer of the *Raspberry Pi* and the development board of the *Smart Pi Expansion Module* SPEM are designed and operated independently, and the communication between them only consumes a part of the ports, the spare ports of the *Raspberry Pi* can be connected with additional hardware to achieve more comprehensive information processing functions.

12.5.3 Practicality

The system used 3.3 V supply to power up the modules. It can supply power to the whole system through the micro-USB port. The developed system had minimal footprint and could be easily connected and separated from the MT electrical cabinet. To meet the long-term working needs, when storing data, it could record the information of each channel in numerical form. By programming, CSV files were automatically stored, which occupy very little storage space. When uploading data, only the last week's data were uploaded, while the database automatically deleted the expired data files.

12.6 FUNCTIONAL STRUCTURE DESIGN OF MONITORING SYSTEM

According to the overall hardware design and requirements proposed in Section 12.2.3.1, this section divides the system functional structure, as shown in Figure 12.2.

The monitoring system was composed of the *Raspberry Pi* information exchange unit and SPEM information acquisition unit. The *Raspberry Pi* interaction unit was composed of three modules: data display module, data storage module, and remote-control module. The SPEM was composed of a signal acquisition module and the data transmission module.

12.6.1 Raspberry Pi Interaction Unit

This section introduces the information exchange unit of the host computer, which is a window for users to obtain intuitive data and information displays. Through this unit, users can further process and analyze the collected data, and judge the state of MTs and tools, so as to realize comprehensive monitoring of MT processing energy consumption. Each function module is introduced as follows:

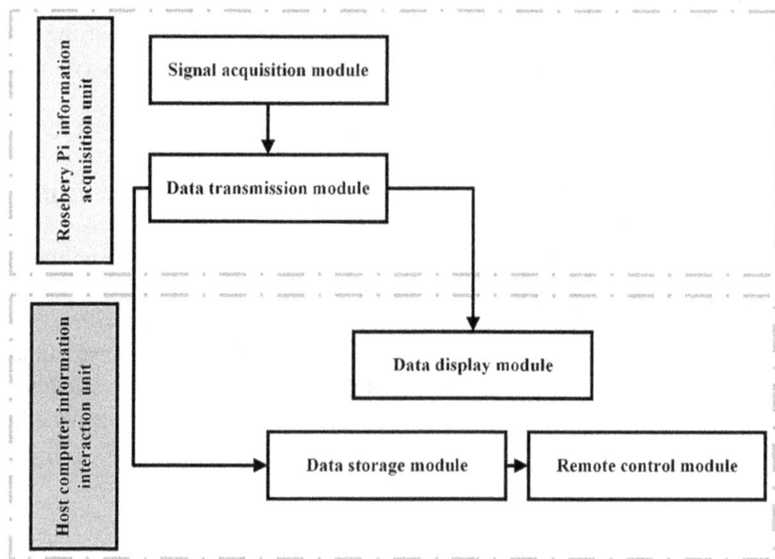

FIGURE 12.2 Schematic diagram of system function structure.

Real-time current signals and real-time power values of each channel were displayed at a refresh frequency of 1 Hz. In addition, the column chart showed the current and the previous day's measurements of the energy consumed by each channel. Users can observe and compare the numerical values of each channel. Real-time power information can also be used as a key parameter to judge the running state of MTs and tool wear. The comparison of real-time information and historical data of energy consumption measurement provides a reference for the formulation of processing schemes and processes from the perspective of energy consumption economy.

The current and power signals collected by the SPEM were stored in the SD memory card of the Raspberry Pi in CSV format. This file records sampling data and time markers at a set frequency, providing data sources for subsequent data analysis and upload.

By setting up the network module of the host computer and loading the corresponding network service protocol and interactive software, remote access and control can be realized. With this module, the monitoring system can be connected and simply fixed in the electrical cabinet of the MT and can log in to the monitoring system remotely under the configuration network environment, and control all the monitoring modules remotely.

12.6.2 SPEM Acquisition Unit

When designing the information acquisition unit of the lower computer, it was needed to consider the sampling frequency of the signal, the analog-to-digital conversion of the signal, the data transmission with the *Raspberry Pi* and the power supply of the mainboard, and then select the appropriate electronic components to cooperate with the sensor and the Raspberry Pi to complete the system hardware platform. The functions of each module of the unit are described as follows:

In the information acquisition unit of the SPEM, the development board as the core part deals with the analog signals collected by sensors and converts the analog signals into digital signals and transmits them to the upper computer. The communication and data transmission between the *Smart Pi Expansion Module* SPEM and the *Raspberry Pi* were carried out by serial bus. In addition, the SPEM has no storage function and an independent power supply interface. It needs efficient data transmission and power supply through a serial bus.

12.7 SELECTION AND DESIGN OF SYSTEM HARDWARE

The monitoring system was designed around three functional modules: signal processing, transmission, and display. In the choice of hardware, the premise is to achieve relevant functions, which is characterized by low cost and lightweight. Hardware includes sensors, *Raspberry Pi*, and lower computer. The performance of these three main electronic components directly determines the performance and performance of the monitoring system.

FIGURE 12.3 Current transformer physical map.

12.7.1 Sensor Selection

This monitoring system needed to monitor the energy consumption information of the whole process of NC MT, so it is needed to measure and record the power distribution with time during the process. Real-time current signals can be obtained by connecting the open-close current transformer with the wires of the components in the electrical cabinet of the NC MT. As shown in Figure 12.3, this system used SCT013-050 open-close current transformer produced by Beijing Yaohua Dechang Electronics Co., Ltd., as a current sensor. The current transformer has a rated input of 50 A, a rated output of 1 V, a sampling frequency range of 50 Hz–1 KHz, and a working temperature of –25~70~C.

To calculate real-time power data, besides the current signal, the voltage signal is also needed. Current and voltage interfaces were reserved in the sensor interface design of the monitoring system. Because laboratories or factories are equipped with high-power regulators, there is little change in voltage during the duration of processing, so it is impossible to analyze the change of energy consumption accordingly. Therefore, the laboratory voltage was measured only before the experiment begins. In the same area, the electrical components were connected to the power grid in parallel, so it is not necessary to directly measure the MT voltage, as long as the incandescent lamp voltage in the same area is measured when the machine is open. The results show that the voltage of the MT in the laboratory is stable, and the maximum change is 8%. In addition, the measurement of voltage required bare wires to be connected to the monitoring system. There are certain risks and uncertainties. The monitoring system may also have an impact on the NC MT itself. Therefore, in subsequent experiments, the voltage was treated as 380 V constant voltage, and the change in machine power was reflected by the current signal measured by the current transformer.

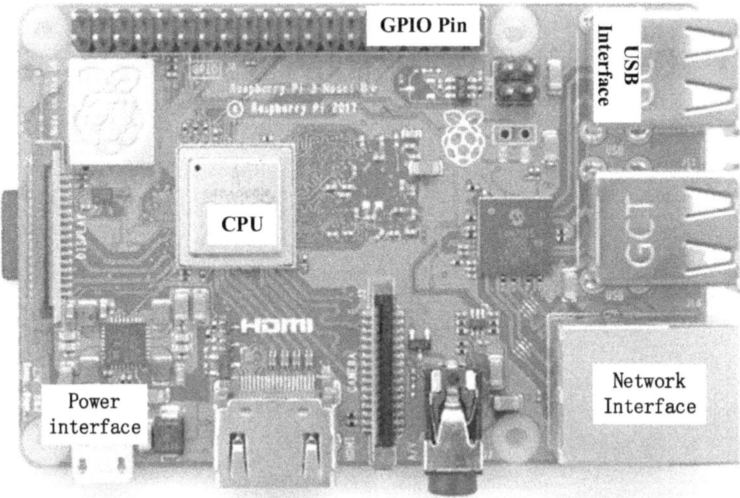

FIGURE 12.4 Pictorial view of raspberry Pi.

12.7.2 Selection of Raspberry Pi

As the core device of information processing and storage, the performance of the host computer is essential for the realization of the main functions of the monitoring system. This system used Raspberry Pi3 Model B+ as the upper computer. As shown in Figure 12.4, the host computer is a microcomputer motherboard based on ARM architecture. It carries 1 GB memory, supports wireless network and Bluetooth 4.1, and has a design size of 56 mm × 85 mm. SD card was used as a memory hard disk, which utilizes considerable computing power. In addition, the most important feature of Raspberry Pi3 is that it is equipped with 40 GPIO pins, which can be connected to other development boards, microcontrollers, or sensors to achieve rich expansion functions.

The *Raspberry Pi* consumes about 10 W energy under 5 V working voltage. Its powerful computing function is the premise of data processing and display. The Bluetooth and wireless network module on board provide the possibility for the realization of remote-control function. The 40 GPIO pins provide the basis for data transmission between upper and lower computers. Specific peripheral configurations and functions are shown in Table 12.1.

12.7.3 Selection of Smart Pi Expansion Module

The SPEM was connected to the sensor for signal acquisition and data processing with the *Raspberry Pi*. This monitoring system used the Smart Pi development board produced by ND GmbH Company, which carries ADE7878 chip, built-in ADC module, and GPIO pin. The development board is the core component of the SPEM information acquisition unit. The transformer connects to the SPEM through three current

TABLE 12.1
Host Computer Peripherals and Functions

Peripheral	Raspberry Pi3
RAM (MB)	1024
CPU	ARM 64, 1.2 GHz
GPIO	40 Pin
Operating voltage (V)	5
Rated power (W)	4 W (5 V/800 mA)
Network interface	802.11n wireless network

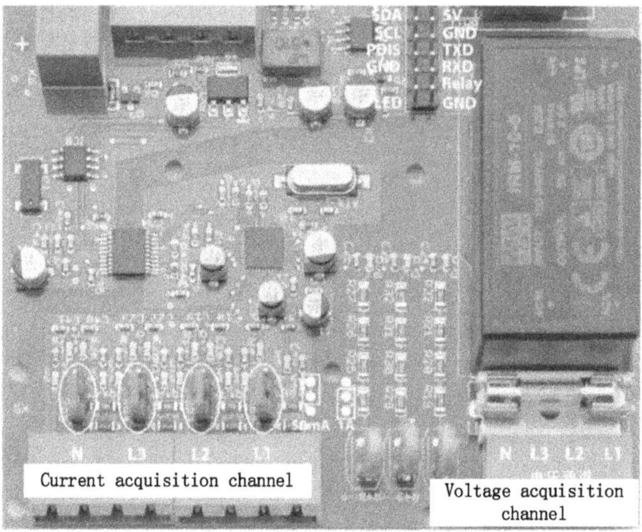

FIGURE 12.5 Top view of SPEM.

channels to realize the signal acquisition and preliminary processing. After signal processing, it is transmitted to the Raspberry Pi for further analysis and processing of data. The physical image of the SPEM is shown in Figure 12.5.

12.8 SYSTEM HARDWARE CIRCUIT DESIGN

To realize the information acquisition function of the lower computer, a crystal oscillator is needed to provide the clock signal with the required frequency. The power supply of the whole monitoring system comes from the micro-USB power interface of the host computer. The pin connection between the two boards completes the power supply of the host computer to the SPEM and realizes the function of data transmission. According to the above requirements, this section designs the crystal oscillator circuit and the board connection circuit and introduces the integrated power supply circuit of the upper computer.

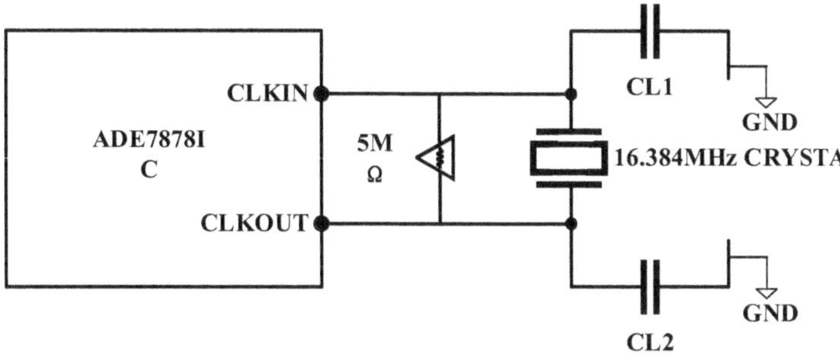

FIGURE 12.6 Crystal circuit design.

12.8.1 Crystal Oscillator Circuit Design

The main body of the information acquisition function of the SPEM is the ADE7878 chip, which provides the waveform sampling data of all three-phase and zero-line currents. Active/reactive power can be collected by configuration. It needs a 3.3 V single power supply and 40 pins to realize the flexible serial interface of I2C, SPI, and HSDC. The signal processing function in the chip is complete, but the serial bus protocol needs a data signal and clock signal. The system provides a clock signal to the chip by an external crystal oscillator. This section introduces the design of a crystal oscillator circuit of 7878 chip.

In the figure, AD7878IC is the chip model, CLKIN and CLKOUT are clock pins, CL1 and CL2 are 24 pF capacitors, and GND is the ground pin. The chip can apply a digital clock signal on the CLKIN pin or a crystal oscillator with a specific frequency. The system uses VM6-1D11C12 16.384 MHz external crystal oscillator, and a 5 M resistance and a 24 pF load capacitor are connected in parallel to ensure the stable operation of the circuit. Pin connection and component selection of the crystal oscillator circuit are shown in Figure 12.6.

12.8.2 Design of Inter-board Connection Circuit

In this system, the pins of the two boards are connected and packaged by DuPont line between the upper and lower boards. The connection between two boards, encapsulation, and key function modules is shown in Figure 12.7.

In the aspect of circuit design, the ADE7878 chip equipped with SPEM carries a fully licensed I2C interface, while Raspberry Pi has no specific serial bus interface. However, the GPIO pin of the host computer contains data I/O pin SDA and serial clock pin SCL, so it is necessary to simulate the IC2 interface to realize the communication between the host computer and the slave computer. Circuit design and pin connection are shown in Figure 12.8.

The circuit schematic diagram (Figure 12.8) shows the pin connection necessary for power supply and data transmission of the system. The pin connection of other functional modules will be shown in the circuit diagram of corresponding modules. From the circuit diagram of interconnection between boards, it can be seen that the SPEM is not connected with power input, and the Raspberry Pin not only supplies

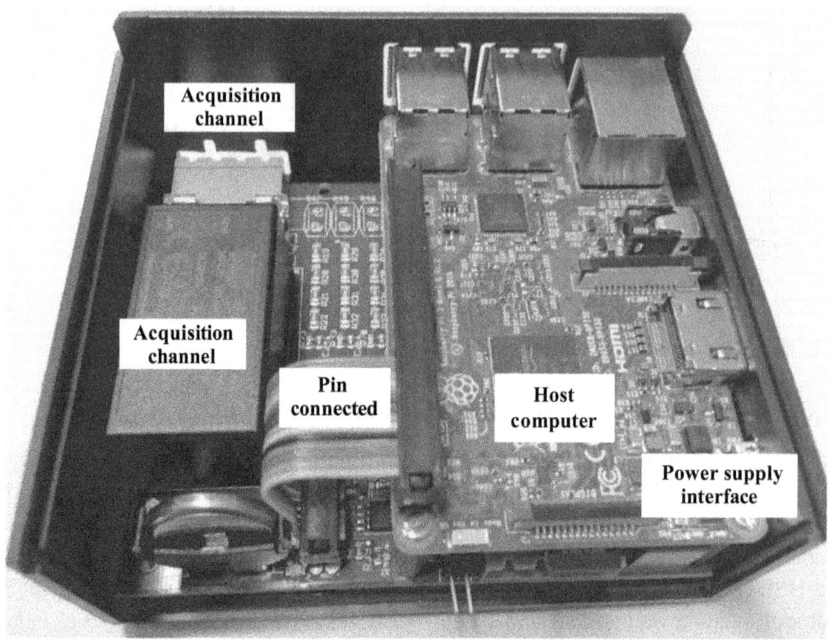

FIGURE 12.7 Physical maps of board connection.

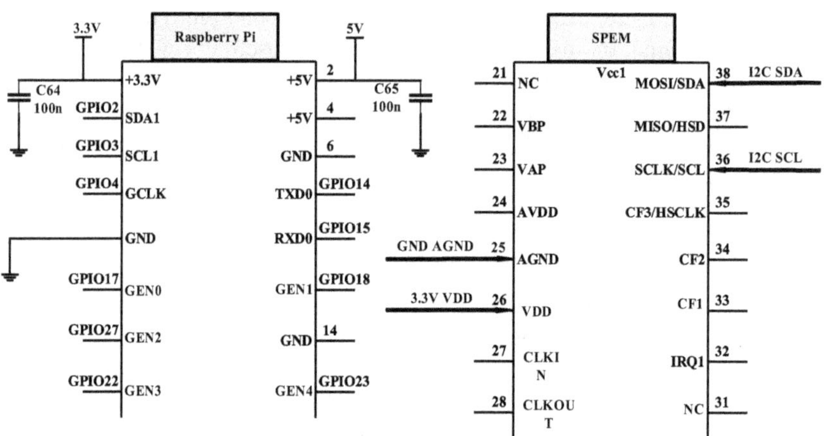

FIGURE 12.8 Raspberry Pi and SPEM connection circuit schematic.

power to itself but also provides power input to the SPEM through the external power supply. The working voltage of the Raspberry Pi is 5 V, and the SPEM supplies power by connecting the 3.3 V pin of the upper computer. The power supply circuit of the Raspberry Pi is shown in Figure 12.9.

Similarly, the grounding pins are common. A voltage-stabilizing circuit is integrated into the upper computer. After the system is connected, the LED indicator

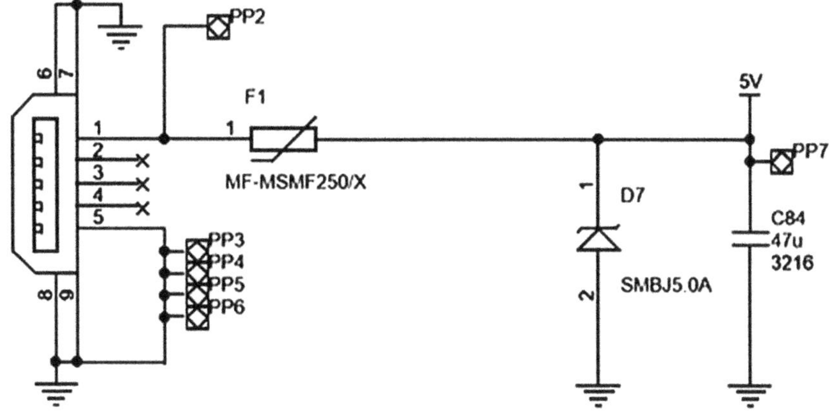

FIGURE 12.9 Host computer power supply circuit diagram.

lights up, indicating that the user's circuit has been energized. In addition, connecting the corresponding SDA and SCL pins to transmit clock signals and data signals can realize data transmission by GPIO pin analog I2C bus.

12.9 THE SOFTWARE OF THE MONITORING SYSTEM

Once enough data are obtained, the following task is about how to get access to data for further processing. Normally, researchers use data acquisition cards and corollary professional software to create an UI.

But, in this case, we have a new viewpoint using low-threshold programming software which just needs a basic understanding of Java script but able to create customized UI. So we turn sight to Node-RED, a flow-based programming tool for the IoT. It provides a browser-based flow editor to wire together flows that also means users can easily get access to data with a browser in mobile devices. Meanwhile, the light-weight design of Node-RED makes it idle for running on low-cost hardware *Raspberry Pi*. Figure 12.10 shows flow built-in Node-RED to record three-phase current signals and store measured data in Raspberry Pi as a CSV file. Moreover, data will upload with the MQTT transmission protocol for remote access.

12.10 VALIDATION OF MEASURED RESULTS

A detailed procedure for power measurement is done on MIKRON UCP 710 vertical milling center. Power consumption at the process level, spindle level, and MT level was measured using the smart meter as shown in Figure 12.11. The current and voltages measured from smart sensors were put in Equation (12.1) to get power and Equation (12.2) was used to measure corresponding energy [16].

$$P = I \times V \times \sqrt{3} \times \cos\varphi \tag{12.1}$$

$$E = P \times t \tag{12.2}$$

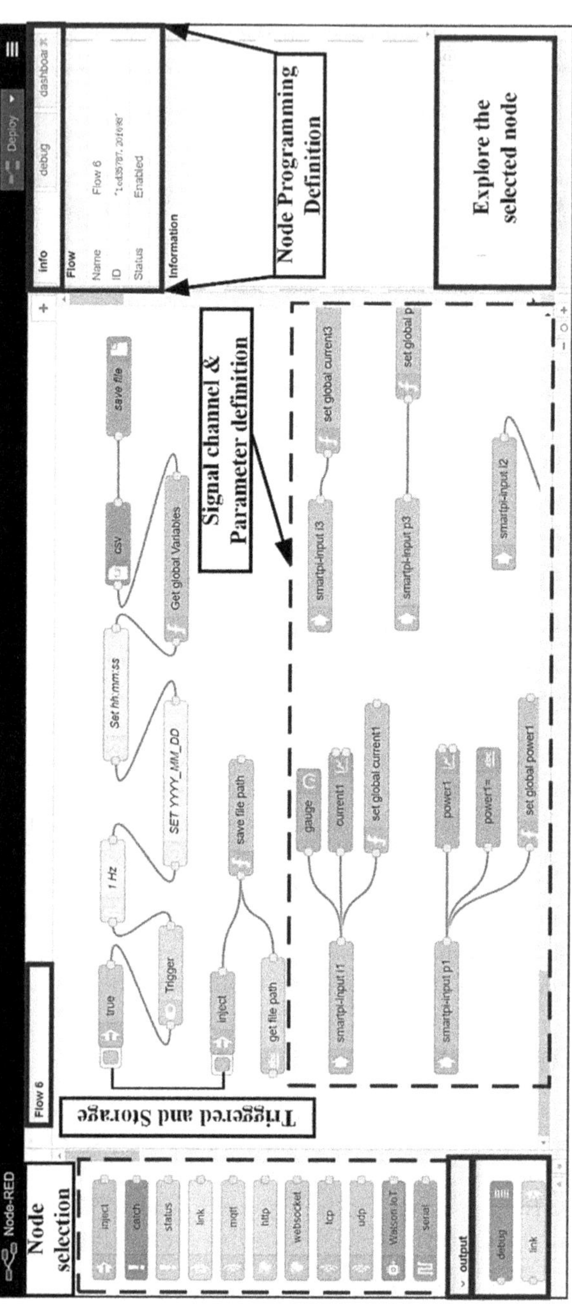

FIGURE 12.10 Software programming interface – Node-RED.

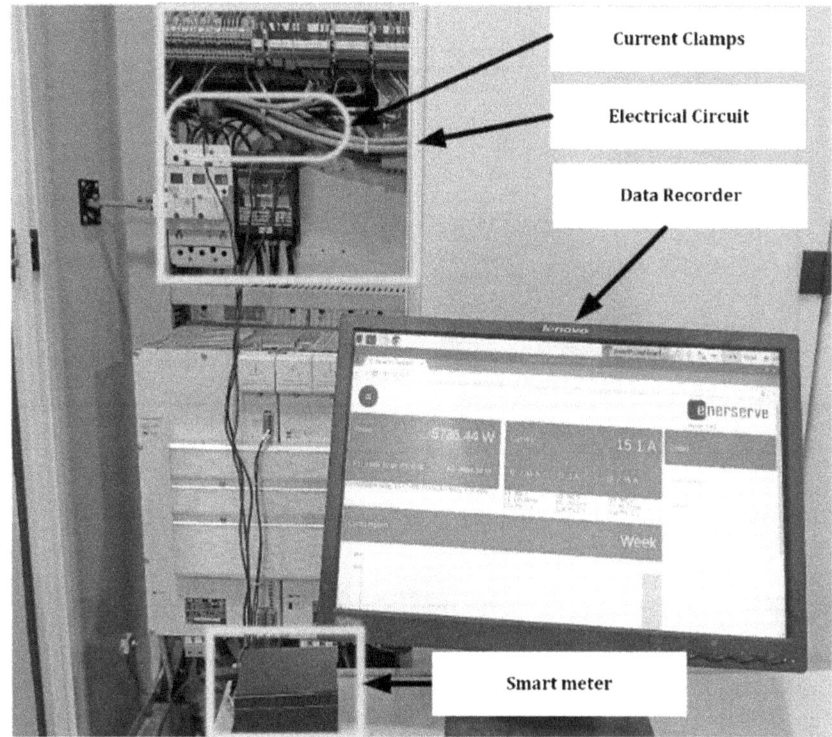

FIGURE 12.11 MT power measurement setup.

Here E, P, and t represent energy consumption, power consumption, and process time in general. The smart meter was fabricated at the laboratory by assembling various components and integrating them with software. It was necessary to verify the measurement of a smart meter by comparing commercially available expensive sensors. The Load Controls Power Meter model (PPC-3) is a commercially available power measuring sensor [15]. Three clip-on voltage sensors work to 600 V. Two isolated analog outputs work on 0–10 V DC and 4–20 milliamp DC. It was procured from Load Controls Incorporated, China. The data of a power sensor can be recorded by Arduino software. The power and energy consumption profile of MT is shown in Figure 12.12.

According to *Machinery's Handbook* [17], as the hardness of workpiece increases less cutting speed is recommended to avoid shorter tool life. Iqbal et al. [18] investigated the effect of increasing yield strength on the cutting energy and found the most significant factor which is responsible for increasing cutting energy. As the hardness of workpiece increases, more cutting forces are required to surpass the resistance of the material to localized plastic deformation. It is a well-known fact that as the cutting forces increase cutting power also increases and the amount of increase in cutting power depends on cutting speed. Theatrically it can be explained as $P_c = F_c \times v$ (Figure 12.13).

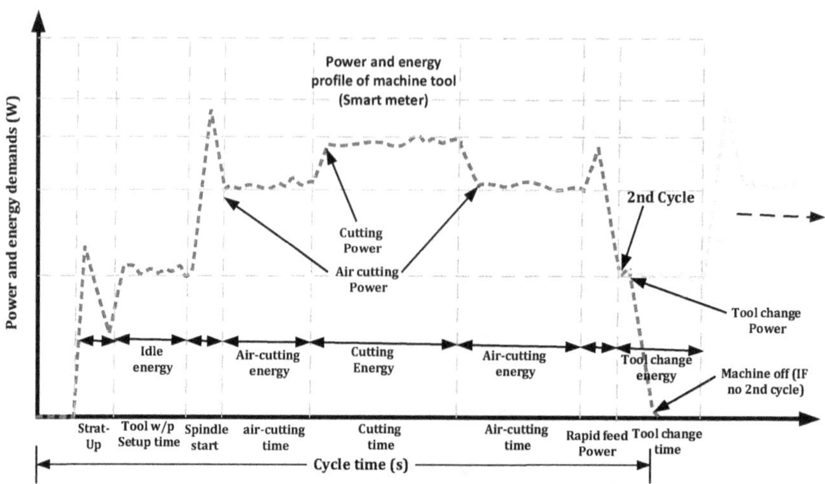

FIGURE 12.12 Power and energy profile of MT measured by the smart meter.

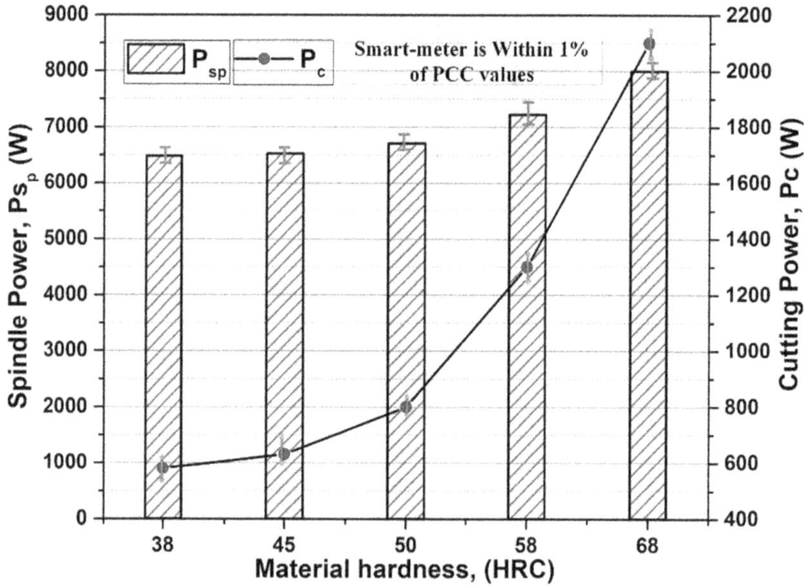

FIGURE 12.13 Effect of material hardness on spindle power and cutting power $\left(v_c = 200 \frac{\text{m}}{\text{min}};\ f_z = 0.06;\ a_p = 5\ \text{mm};\ a_e = 1.5\ \text{mm} \right)$.

The several validation runs were performed on MIKRON UCP 710 vertical milling center. The spindle of the MT was rotated at various RPM to measure the varying power. Figure 12.14 illustrates the power measurement by smart meter was within 1% variation of power measured by PPC-3 power measurement meter. However,

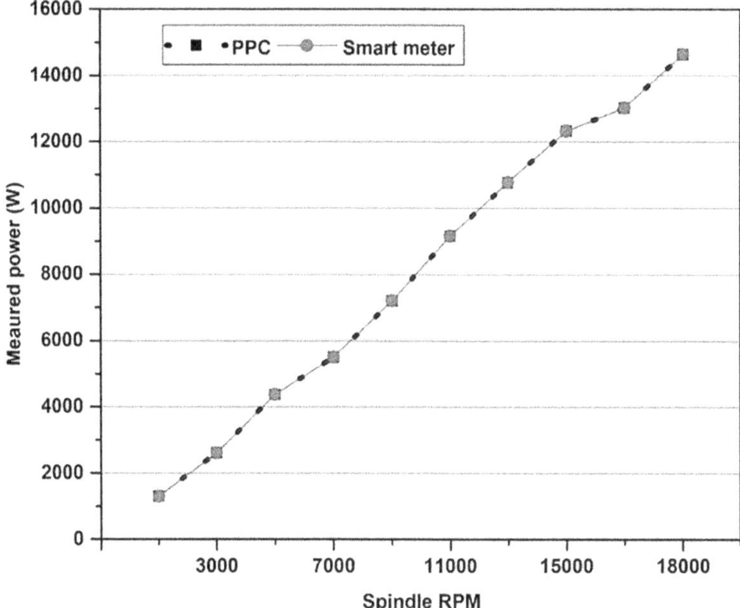

FIGURE 12.14 Validation of the measured data.

data storage and analysis in PPC-3 sensors are more versatile and some more efforts are needed to enhance the performance of smart meter by updating its software programming.

The main limitation of the fabricated sensor is the data storage which can be solved by updating the software. The efforts are needed to integrate the smart meter with the CNC panel to get real-time data.

12.11 CONCLUSION

Aiming at the requirement of MT energy consumption monitoring and related functional modules, this chapter designs the overall scheme of the monitoring system and builds the hardware platform of the system based on the requirements of the overall design. The main outcomes are as follows:

1. The energy consumption is analyzed from two aspects: the composition of energy consumption components and the change of MT processing state. Through the analysis of the former, the energy consumption structure of the MT itself is defined, and the main target of the system monitoring, namely the variable energy consumption components of the MT, is determined.
2. In the overall design of the monitoring system, the objectives and principles of the system design are determined, and the working principle of the system is elaborated. On this basis, the system is divided into two units, including the lower computer information acquisition unit and the upper

computer information exchange unit. At the same time, the internal modules of each unit are designed. The lower computer information acquisition unit is divided into two modules: information acquisition and data transmission. The upper computer information exchange unit is divided into three units: data display, data storage, and remote control. The functions of each module are defined.

3. According to the overall design requirements of the MT processing energy consumption monitoring system, key electronic components are selected, including SCT013-050 open-close current transformer, Raspberry Pi PC motherboard and Smart Pi development board, and the overall structure of the system is determined.
4. According to the hardware characteristics of the selected electronic components, the crystal oscillator circuit is designed to provide the clock signal needed for signal acquisition, and the board connection circuit is designed to realize the power supply function of the lower computer and the data transmission channel between the two boards, which provides the hardware basis for the software development of the data transmission module.
5. The performance of the smart meter was validated, and results were found in good agreement with the commercially available sensor.

ACKNOWLEDGMENT AND FUNDING

This work was supported by the U1601204 the National Key Research and Development Project (Grant No. 2018YFB2002202).

REFERENCES

1. May G, Barletta I, Stahl B, Taisch M (2015) Energy management in production: A novel method to develop key performance indicators for improving energy efficiency. *Appl Energy,* 149:46–61.
2. Feng L, Mears L, Beaufort C, Schulte J (2016) Energy, economy, and environment analysis and optimization on manufacturing plant energy supply system. *Energy Convers Manag* 117(7):454–465.
3. China NB of S of (2018) *China Energy Statistical Yearbook: Energy Consumption*, China Statistics Press, China.
4. EIA (2018) Annual Energy Outlook 2018-EIA. https://www.eia.gov/outlooks/aeo/pdf/AEO2018.pdf. Accessed 16 Nov 2016
5. Chan DYL, Huang CF, Lin WC, Hong GB (2014) Energy efficiency benchmarking of energy-intensive industries in Taiwan. *Energy Convers Manag* 77:216–220.
6. Khan AM, Jamil M, Salonitis K, et al. (2019) Multi-objective optimization of energy consumption and surface quality in nanofluid SQCl assisted face milling. *Energies* 12:710.
7. NBS (2019) National Bureau of Statistics of China. Nanjing.
8. Cai W, Liu F, Dinolov O, et al. (2018) Energy benchmarking rules in machining systems. *Energy* 142:258–263.
9. Merchant ME (1945) Mechanics of the metal cutting process. I. Orthogonal cutting and a type 2 chip. *J Appl Phys* 16:26, https://doi.org/10.1063/1.1707586.

10. Ertunc HM, Loparo KA, Ocak H (2001) Tool wear condition monitoring in drilling operations using hidden Markov models (HMMs). *Int J Mach Tools Manuf.* 41(9):1363–1384.
11. Abou-El-Hossein KA, Kadirgama K, Hamdi M, Benyounis KY (2007) Prediction of cutting force in end-milling operation of modified AISI P20 tool steel. *J Mater Process Technol.* 182(1–3):241–247.
12. Dahmus JB, Gutowski TG (2004) *An Environmental Analysis of Machining.* In: American Society of Mechanical Engineers, Manufacturing Engineering Division, MED Paper No: IMECE2004-62600, pp. 643–652; 10 pages, https://doi.org/10.1115/IMECE2004-62600.
13. Li L, Yan J, Xing Z (2013) Energy requirements evaluation of milling machines based on thermal equilibrium and empirical modelling. *J Clean Prod* 52:113–121.
14. Lv JX, Tang RZ, Jia S (2012) Methodology for calculating energy consumption of a machining process. In: *Advanced Materials Research* Vols. 472–475:2736–2743.
15. Zhao Z, Fu Y, Xu J, et al. (2016) An investigation on high-efficiency profile grinding of directional solidified nickel-based superalloys DZ125 with electroplated CBN wheel. *Int J Adv Manuf Technol* 83:1–11.
16. Edem IF, Balogun VA, Mativenga PT (2017) An investigation on the impact of toolpath strategies and machine tool axes configurations on electrical energy demand in mechanical machining. *Int J Adv Manuf Technol* 92:2503–2509.
17. Oberg E, Jones FD, Horton HL AHHR (2012) *Machinery's Handbook: A Reference Book for the Mechanical Engineer, Designer, Manufacturing Engineer, Draftsman, Toolmaker, and Machinist* Industrial Press, New York.
18. Iqbal A, Biermann D, Abbas H, et al. (2018) Machining β-titanium alloy under carbon dioxide snow and micro-lubrication: A study on tool deflection, energy consumption, and tool damage. *Int J Adv Manuf Technol* 97:4195–4208.

Section 4

Modeling and Simulation

Modeling and simulation are important tools for gauging the performance of real system before these are manufactured and put into operation. The two chapters in this section are the replication of real systems using modern modeling and simulation tools.

13 A Six Degree of Freedom Machining Bed

Kinematic Model Development, Verification, and Validation

Muhammad Faizan Shah and Kamran Nazeer
Khwaja Fareed University of Engineering and IT

Zareena Kausar
Air University

Muhammad Umer Farooq, Syed Saad Farooq, and Ghias Mahmood Khan
Khwaja Fareed University of Engineering and IT

CONTENTS

13.1 INTRODUCTION

Removal of material from a workpiece to obtain the desired shape is termed as machining. Milling, turning, and drilling are statistically the most common machining operations. Machining is a part of manufacturing metal products, but it is also used in the processing of various other metals such as ceramics, plastics, and wood. Sawing, broaching, boring, and shaping are varied categories of machining. Examining machining processes closely divulges that a major holdup in reducing machining lead-time is the prolonged workpiece setup time. Spindle technologies used currently allow quick material removal due to their high-power ratings and high rotational speeds. This high-speed material removal reduces operation time, but unfortunately, it also compromises the performance envelope of high spindle machines. These high spindle tools are highly vulnerable to external and internal vibrations. These vibrations are a major cause in affecting the accuracy and precision of the desired output. A solution to these problems is to restrict the motion of the tool and give the workpiece motion in six degrees of freedom. For this purpose, a six degree of freedom parallel manipulator is proposed [1].

The purpose of proposing a parallel manipulator for this problem is that it offers high dynamics, high stiffness, and offers less positional accuracy errors [2]. Nowadays parallel manipulators have gained a lot of attention from the industrial sector due to their ability to bear high load, high control, and higher accuracy [3]. Apart from these attributes of parallel manipulators these manipulators have some drawbacks of difficult forward kinematics and smaller workspace [4–8]. In general, forward kinematics of parallel manipulators is difficult to solve [9–14].

The parallel manipulator proposed in this chapter is a Stewart platform in which the tool is fixed and all the motion is generated by the bed on which workpiece is lying [15]. With the flexibility of parallel manipulators, the tool may not be required to be replaced again and again. In traditional machining operations, the whole setup is shut down to replace the tool and hence increasing the setup time. The proposed parallel manipulator can give six degrees of freedom to the workpiece which includes three lateral motions and three angular motions.

There are many configurations of parallel manipulators. The configuration proposed in this chapter is the 3–6 configuration of a Stewart platform. The proposed machining bed can be used for milling, drilling, and cutting operations. In the configuration the base plate is divided into three sections; in each section two kinematic links are joined parallel adjacent to each other. The schematic of the proposed parallel machining bed is shown in Figure 13.1.

13.2 KINEMATIC MODEL

All legs of the parallel manipulating machining bed were selected to be identical. A kinematic model of one leg was, therefore, developed which was then integrated over the number of legs, i.e., six. Figure 13.2 shows a schematic view of the leg (an *i*th leg) of the machining bed. T_{P_i} refers to the anchor point of the top plate and B_{P_i} refers to the anchor point of the base plate. The base plate was taken to be a reference frame. *Len* represents the leg length of the *i*th leg. TV is the translation vector and H_i is the resultant vector that was obtained from the head to tail rule of vector addition. Various orientations were proposed by the researchers for this type of parallel manipulators.

$$\mathrm{Len}_i = \mathrm{TV} + {}^{P}_{B}RT_{Pi} - B_{Pi} \tag{13.1}$$

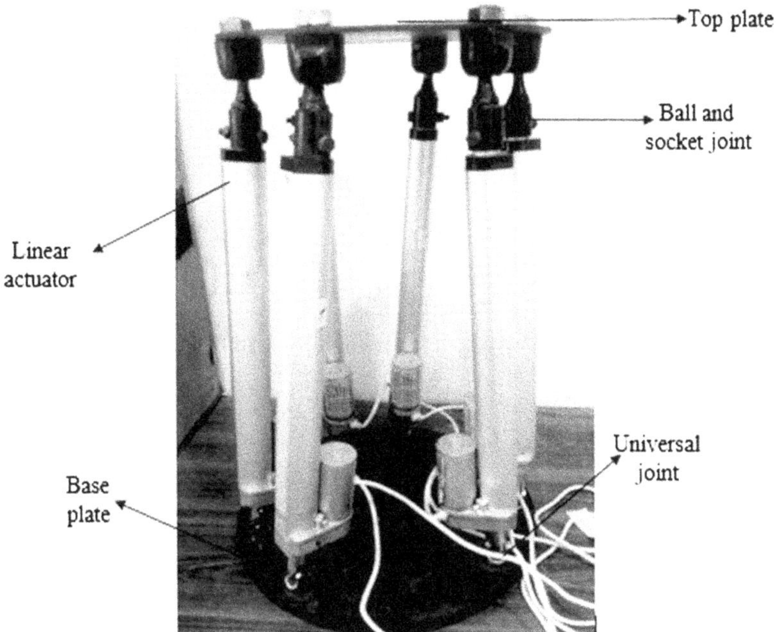

FIGURE 13.1 The proposed parallel machining bed.

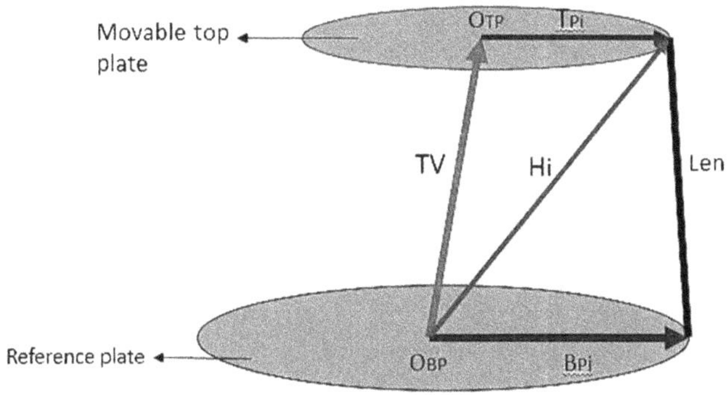

FIGURE 13.2 Schematic diagram of an *i*th leg of proposed machining bed.

Equation (13.1) gives the leg length of an *i*th leg using geometric relations and vector addition method.

where R is the rotation matrix from O_{BP} to O_{TP} and was calculated using Equations (13.2) and (13.3).

$$ {}_{B}^{R}R = R_Z(\psi) + R_Y(\theta) + R_X(\varphi) \tag{13.2}$$

In Equation (13.3) c represents cosine and s represents sine of angles, whereas ψ, θ, and φ are angles in respective axis.

$$ {}_{B}^{R}R = \begin{pmatrix} c(\psi)c(\theta) & -s(\psi)c(\varphi)+c(\psi)s(\theta)s(\varphi) & s(\psi)s(\varphi)+c(\psi)s(\theta)c(\varphi) \\ s(\psi)c(\theta) & c(\psi)c(\varphi)+s(\psi)s(\varphi)s(\theta) & -c(\psi)s(\varphi)+s(\psi)s(\theta)c(\varphi) \\ -s(\theta) & c(\theta)s(\varphi) & c(\theta)c(\varphi) \end{pmatrix} \tag{13.3} $$

Figure 13.3 shows the distribution of ball and socket joints on the top plate of the proposed machining bed. The angle between each joint is 60°. Figure 13.3 also shows the distribution of universal joints on the base plate. The universal joints were attached in a pair of two at the base plate with an angle of 35° between them and an angle of 85° between each pair.

Equations (13.4)–(13.9) were formulated for the position of joints at the top plate, where "r" is the radius of the top plate.

$$ T_{P1} = \begin{bmatrix} r & 0 & 0 \end{bmatrix} \tag{13.4} $$

$$ T_{P2} = \begin{bmatrix} r\cos 60 & r\sin 60 & 0 \end{bmatrix} \tag{13.5} $$

$$ T_{P3} = \begin{bmatrix} -r\cos 60 & r\sin 60 & 0 \end{bmatrix} \tag{13.6} $$

$$ T_{P4} = \begin{bmatrix} -r & 0 & 0 \end{bmatrix} \tag{13.7} $$

$$ T_{P5} = \begin{bmatrix} -r\cos 60 & -r\sin 60 & 0 \end{bmatrix} \tag{13.8} $$

$$ T_{P6} = \begin{bmatrix} r\cos 60 & -r\sin 60 & 0 \end{bmatrix} \tag{13.9} $$

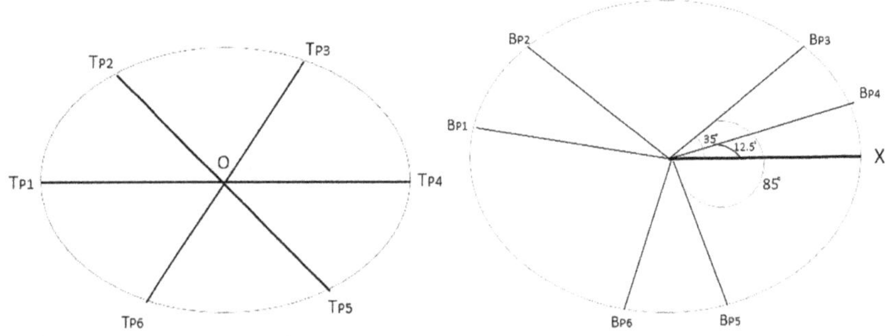

FIGURE 13.3 Angle distribution on movable top plate and fixed base plate.

Equations (13.10)–(13.15) were formulated for the position of joints at the base plate, where "R" is the radius of the base plate.

$$B_{P1} = \begin{bmatrix} R\cos 12.5 & R\sin 12.5 & 0 \end{bmatrix} \tag{13.10}$$

$$B_{P2} = \begin{bmatrix} R\cos 47.5 & R\sin 47.5 & 0 \end{bmatrix} \tag{13.11}$$

$$B_{P3} = \begin{bmatrix} -R\cos 47.5 & R\sin 47.5 & 0 \end{bmatrix} \tag{13.12}$$

$$B_{P4} = \begin{bmatrix} -R\cos 12.5 & R\sin 12.5 & 0 \end{bmatrix} \tag{13.13}$$

$$B_{P5} = \begin{bmatrix} -R\cos 17.5 & -R\sin 17.5 & 0 \end{bmatrix} \tag{13.14}$$

$$B_{P6} = \begin{bmatrix} R\cos 17.5 & -R\sin 17.5 & 0 \end{bmatrix} \tag{13.15}$$

Using the equations obtained from Figure 13.3 and Equation (13.1) obtained for leg length, the solution was obtained for each leg length of the machining bed. Equations (13.16)–(13.21) show that each pair of any two legs have the same pattern of leg motion.

$$\text{Len}_1 = \begin{pmatrix} rc\psi c\theta - Rc12.5 \\ rs\psi c\theta - Rc12.5 \\ a - rs\theta \end{pmatrix} \tag{13.16}$$

$$\text{Len}_2 = \begin{pmatrix} rc60c\psi c\theta - rs60s\psi s\varphi + rs60c\psi s\theta s\varphi - Rc47.5 \\ rc60s\psi c\theta + rs60c\psi c\varphi + rs60s\varphi s\psi s\theta - Rs47.5 \\ a - rc60s\theta + rs60c\theta s\varphi \end{pmatrix} \tag{13.17}$$

$$\text{Len}_3 = \begin{pmatrix} -rc60c\psi c\theta - rs60s\psi s\varphi + rs60c\psi s\theta s\varphi + Rc47.5 \\ -rc60s\psi c\theta + rs60c\psi c\varphi + rs60s\varphi s\psi s\theta - Rs47.5 \\ a + rc60s\theta + rs60c\theta s\varphi \end{pmatrix} \tag{13.18}$$

$$\text{Len}_4 = \begin{pmatrix} -rc\psi c\theta + Rc12.5 \\ -rs\psi c\theta - Rs12.5 \\ a + rs\theta \end{pmatrix} \tag{13.19}$$

$$\text{Len}_5 = \begin{pmatrix} -rc60c\psi c\theta + rs60s\psi c\varphi + rs60c\psi s\theta s\varphi + Rc17.5 \\ -rc60s\psi c\theta - rs60c\psi c60 - rs60s\varphi s\psi s\theta + Rs17.5 \\ a + rc60s\theta - rs60c\theta s\varphi \end{pmatrix} \tag{13.20}$$

$$Len_6 = \begin{pmatrix} rc60c\psi c\theta + rs60s\psi c\varphi - rs60c\psi s\theta s\varphi - Rc17.5 \\ rc60s\psi c\theta - rs60c\psi c60 - rs60s\varphi s\psi s\theta + Rs17.5 \\ a - rc60s\theta - rs60c\theta s\varphi \end{pmatrix} \tag{13.21}$$

13.2.1 Case Study

The kinematic model presented above has the benefit of yielding two same leg lengths for a machining position. A code was developed in MATLAB® to verify the kinematic model. Various positions were given as an input and leg lengths of the proposed machining bed were obtained as an output. The leg lengths obtained are given in Table 13.1.

From Table 13.1, the proposed hypothesis of obtaining two same leg lengths for a position was verified. This results in less computation time.

13.3 VERIFICATION OF KINEMATIC MODEL

To verify the kinematic model obtained in Equations (13.16–13.21) of the previous section, a number of various simulation tests were carried out for an E-shaped workpiece shown in Figure 13.4. The workpiece to be machined consisted of five lines out of which three were horizontal and two were vertical. The machining test to be performed was divided into six tests. The first test was to move the workpiece from

TABLE 13.1
Leg Lengths Obtained for Case Study

Position	Leg	Length (mm)
[0; 100; 380]	Len1	397.38
	Len2	458.10
	Len3	458.10
	Len4	397.38
	Len5	404.37
	Len6	404.37
[100; 50; 380]	Len1	389.29
	Len2	505.32
	Len3	505.32
	Len4	409.18
	Len5	445.65
	Len6	384.96
[40; 100; 380]	Len1	395.40
	Len2	479.05
	Len3	479.05
	Len4	403.35
	Len5	414.70
	Len6	389.64

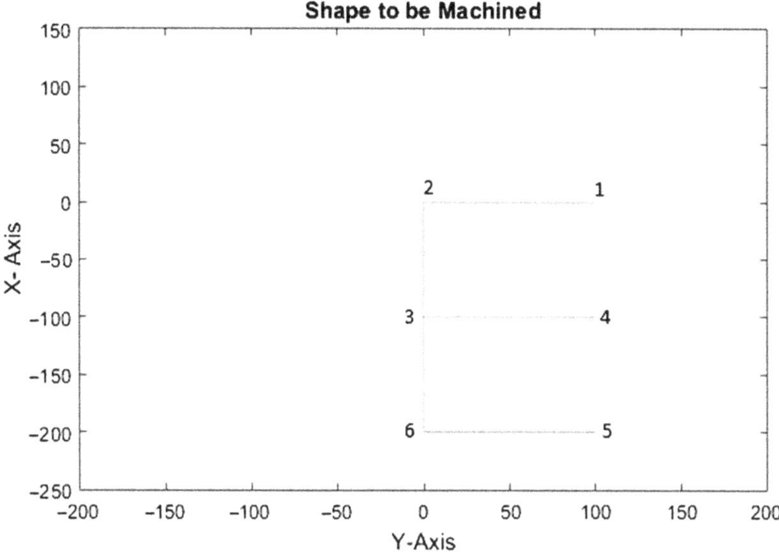

FIGURE 13.4 The desired shape of the workpiece to be machined.

points 1 to 2 as shown in Figure 13.4. Coordinates for starting point were set to be [0; 100; 380] according to reference axis taken. The second test was to move the workpiece from point 2 to 3. Coordinates for starting point were set to be [0; 0; 380]. The third test was to move the workpiece from point 3 to 4 in a horizontal position. Coordinates for starting point were set to be [100; 0; 380]. The fourth test was to move the workpiece from point 4 to 5 but without any contact between the tool and workpiece. The fifth test was to move the workpiece from point 5 to 6 in the horizontal position. Coordinates for starting point were set to be [200; 100; 380]. Sixth and last test was to move the workpiece vertically from point 6 to 3. Coordinates for starting point were set to be [200; 0; 380]. From Figure 13.5 the leg lengths obtained for these tests can be seen.

The machining process to be carried out on the workpiece was divided into six tests. Tests 1, 3, and 5 were done to perform machining on the workpiece when the bed was translating in Y-axis direction. There was no translation in X-axis during this test. Tests 2, 4, and 6 were carried out to perform machining on the workpiece in Y-axis direction. In this test, the translation was only in Y-axis and no translation was observed on X-axis. In this test, the movement of the machining bed was only in Y-axis. It can be seen from Figure 13.5 that out of six legs length each of the two legs has the same length values.

13.4 EXPERIMENTAL VERIFICATION

The proposed machining bed was tested experimentally to verify the results obtained in Sections 13.2 and 13.3 of this chapter. The physically developed parallel machining bed is shown in Figure 13.1. The workpiece was attached to the

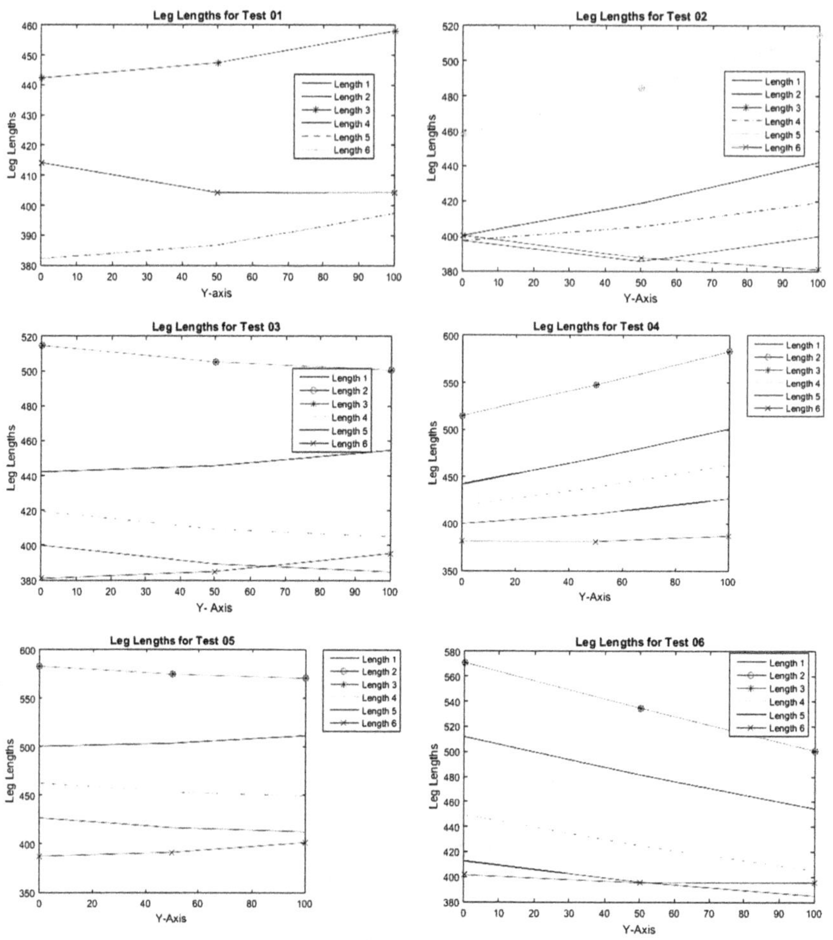

FIGURE 13.5 Lengths of six legs obtained for tests.

movable top plate of the machining bed and the tool was fixed in a chuck. A dot punch tool was converted into a fine cutting tool. For the reference, an E shape was first made from a marker on the workpiece. Reference shape was drawn to check the accuracy and precision of the cut being made because of the motion given to the workpiece. The tool and workpiece movements observed during experiments are shown in Figure 13.6.

Figure 13.7 shows the machined workpiece. The cuts obtained on the workpiece from the tool deviate from the reference shape drawn from the pen. The maximum inaccuracy measured for the complete shape was 5% of the reference position. This slight inaccurate and imprecise cut may be resulted due to error in tool positioning or error in workpiece positioning on the movable top plate or due to both. Play in the joints of the machining bed may also be one of the reasons for the inaccurate and imprecise cut.

FIGURE 13.6 Experimental testing of parallel machining bed.

FIGURE 13.7 Machined work piece.

13.5 CONCLUSION

A new mechanism with six degrees of freedom was proposed as a machining bed. The kinematic model for the proposed bed was derived which was then verified using MATLAB®. The results obtained from MATLAB® support the hypothesis derived from observing the kinematic model of the proposed configuration. The hypothesis was that any two leg lengths of the platform will have the same value. This feature of the configuration decreases computation time and complexity of the motion/positioning of the workpiece for machining of any contour on the workpiece. The experimental setup showed the physical usage of the proposed machining bed. The proposed machining bed can be used for milling, drilling, and cutting operations. Factors affecting the accuracy of the machining need to be studied in the future.

REFERENCES

1. Z. Kausar, M. A. Irshad, and S. Shahid, "A parallel robotic mechanism replacing a machine bed for micro-machining," in *Scientific Cooperations International Workshops on Electrical and Computer Engineering Subfields*, 2014, no. August, pp. 228–233, 2014.
2. STEWART and D., "A platform with six degrees of freedom," *Proc. Instn. Mech. Engrs. Pt. 1*, vol. 180, no. 15, p. 371, 1966.
3. S. Lee, J. Song, W. Choi, and D. Hong, "Position control of a Stewart platform using inverse dynamics control with approximate dynamics," *Mechatronics*, vol. 13, pp. 605–619, 2003.
4. J. Borras, F. Thomas, and C. Torras, "New geometric approaches to the singularity analysis of parallel platforms," *IEEE/ASME Trans. Mechatronics*, vol. 19, no. 2, pp. 173–180, 2014.
5. Q. Jiang and C. M. Gosselin, "Determination of the maximal singularity-free orientation workspace for the Gough-Stewart platform," *Mech. Mach. Theory*, vol. 44, no. 6, pp. 1281–1293, 2009.
6. S. Pedrammehr, M. Mahboubkhah, and N. Khani, "A study on vibration of Stewart platform-based machine tool table," *Int. J. Adv. Manuf. Technol.*, vol. 65, no. 5, pp. 991–1007, 2013.
7. M. F. Shah, Z. Kausar, and F. K. Durrani, "Design, modeling and simulation of six degree of freedom machining bed," *Proc. Pakistan Acadmey Sci.*, vol. 53, no. 2, pp. 163–176, 2016.
8. S. Staicu, "Dynamics of the 6-6 Stewart parallel manipulator," *Robot. Comput. Integr. Manuf.*, vol. 27, no. 1, pp. 212–220, 2011.
9. S. Zarkandi, "Kinematic and dynamic modeling of a planar parallel manipulator served as CNC tool holder," *Int. J. Dyn. Control*, no. January, vol. 6, pp. 14–28, 2018.
10. C. Innocenti and V. Parenti-Castelli, "Direct position analysis of the Stewart platform mechanism," *Mech. Mach. Theory*, vol. 25, no. 6, pp. 611–621, 1990.
11. S. Staicu and D. Zhang, "Dynamic modelling of a 4-DOF parallel kinematic machine with revolute actuators," *Int. J. Manuf. Res.*, vol. 3, no. 2, pp. 172–187, 2008.
12. K. Harib and K. Srinivasan, "Kinematic and dynamic analysis of Stewart platform-based machine tool structures," *Robotica*, vol. 21, no. 5, pp. 541–554, 2003.
13. H. Hajimirzaalian, H. Moosavi, and M. Massah, "Analyzing and simulating the inverse and the direct dynamics of parallel robot Stewart platform," *2nd International Conference on Computer and Network Technology (ICCNT 2010)*, vol. 5, pp. 136–141, 2010.
14. H. Guo and H. Li, "Dynamic analysis of a Stewart platform manipulator," vol. 10, no. 5, pp. 11–20, 2007.
15. M. F. Shah, Z. Kausar, F. K. Durrani, A. H. Tahir, and N. Khalid, "Design and development of six degrees of freedom parallel mechanism for the purpose of machining," in *Student Research Paper Conference*, vol. 2, pp 1–6, 2016.

14 Learning Fault-Tolerant Control Using a Table Sat Platform

Ali Nasir
University of Central Punjab

Ella M. Atkins
University of Michigan Ann Arbor

CONTENTS

14.1 INTRODUCTION

Learning mathematical equations and expressions that represent physical phenomenon is often not intuitive. Even the physical behaviors of the systems themselves are less intuitive quite a few times. But since seeing is believing, it is easier for a student to try to understand and digest in his brain what is observed. On the other hand, digesting mathematics is not so easy.

Table Sat is one of those tools that are used for real-time illustration of how the basic control systems theory works [1,2]. It is specifically designed for learning and implementing feedback control theory but is not limited to it [3]. A student can learn a lot about real-time behavior of sensors and the properties of their noise along with how they are used for feeding back the outputs from the system into the controller. The main sensors embedded in the system include a single-axis gyroscope, a three-axes magnetometer, and four-coarse sun sensors. Table Sat offers two computer fans as actuators that can rotate the system about the z-axis (perpendicular to the surface of the Table Sat) in both clockwise and counterclockwise directions.

It has a Prometheus processor and a wireless modem for communication with the server computer. Power is supplied by a 19 V battery. Figure 14.1 shows the Table Sat platform with all its hardware.

This chapter illustrates how Table Sat can be used to learn feedback control. It will further be illustrated on how to implement the state estimation using Kalman filtering and implementation of a technique for sensor fault identification to indicate the possible use of Table Sat in learning Fault-Tolerant Control. Fault-tolerant control has been studied extensively by the research community. Useful reviews on fault-tolerant control, in general, are found in Refs. [4] and [5] whereas the fault-tolerant control specific for satellite missions is studied in Refs. [6-8]. Specifically, in Refs. [6] and [8] provide Markov Decision Process-based approaches that consider the science mission while accounting for the failure modes. A general fault tolerance framework has been discussed in Ref. [7] for postfault attitude management of spacecraft. Despite the cited work, there has been little effort made in terms of writing textbooks or book chapters that are easy to understand for the students. Therefore, this chapter adds to

FIGURE 14.1 Table Sat platform 0.

the educational resources available to the students for learning fault-tolerant control for satellite missions.

14.2 TABLE SAT MODEL

Table Sat can be modeled approximately as a second-order (double integrator) single-input single- (or multiple) output linear system, although variable friction and platform wobble can become nontrivial effects. The linearized model is presented as follows:

$$\begin{bmatrix} \dot{\theta} \\ \dot{\omega} \end{bmatrix} = \begin{bmatrix} 0 & 1 \\ 0 & 0 \end{bmatrix} \begin{bmatrix} \theta \\ \omega \end{bmatrix} + \begin{bmatrix} 0 \\ 1 \end{bmatrix} u_{\text{fan}} + \begin{bmatrix} 0 \\ 1 \end{bmatrix} w \tag{14.1}$$

$$y = \begin{bmatrix} 0 & 1 \end{bmatrix} \begin{bmatrix} \theta \\ \omega \end{bmatrix} \tag{14.2}$$

Here θ is the rotation angle (or heading) of Table Sat, ω is the angular rate (or angular velocity), u_{fan} is the control torque generated by the pair of fans, and w is the disturbance or friction torque. Equations (14.1) and (14.2) represent a single-input single-output linear system with each value of θ as an equilibrium point and it is open-loop unstable at every single equilibrium point. Further, it is fully controllable through u_{fan} and is unobservable using the output as in Equation (14.2). All these properties can be verified through physical examination of the behavior of Table Sat. For example, if Table Sat is left alone at any heading, it will stay there forever until an external force is applied (thus each value of theta is an equilibrium point). If the Table Sat is disturbed from a heading angle, it will not come back to that angle by itself rather will keep moving until some force causes it to stop (instability). Using the value of the angular velocity of the Table Sat, one cannot tell its heading (unobservability) without knowing the initial value of heading. Finally, using the fans, one can move the Table Sat to any desired value of θ (or ω) in a finite time, starting from any initial heading (controllability).

14.3 VELOCITY CONTROL

For controlling the velocity, a simple PD controller can be used in Refs. [9,10] with the proportional gain set at zero (since heading is not of interest):

$$K = \begin{bmatrix} 0 & \alpha \end{bmatrix} \begin{bmatrix} \theta - \theta_{\text{des}} \\ \omega - \omega_{\text{des}} \end{bmatrix} : (\alpha > 0) \tag{14.3}$$

It turns out that the controller in (14.3) is always turned on trying to achieve the perfect desired velocity. The reason why it cannot achieve the perfect desired velocity is because there is noise in the velocity sensor (the gyroscope). The best this controller can do is that it can keep the Table Sat with-in the region (characterized by sensor noise) around desired angular velocity. Therefore, the controller is defined to be as in Equation (14.3) if $\left(|\omega - \omega_{\text{des}}| > \beta\right)$ and zero otherwise, where β is a constant (in this case 0.1) that indicates uncertainty in sensor readings. The figure below shows some plots obtained from conducting a velocity control experiment on Table Sat where the goal was to ramp up the velocity to the desired value using full throttle and then stay at the desired velocity for 30 seconds before finally turning the fans off and letting the Table Sat to ramp down to zero velocity.

Figure 14.2 shows the results of applying velocity control on Table Sat. The output from the gyroscope is visible to be between 0 and 5 V, with 0 indicating maximum counterclockwise angular velocity and 5 indicating maximum clockwise angular velocity. Zero angular velocity is indicated by 2.353 V in the figure. On the other hand, thrust has a range of −10 to 10 where 10 indicates fan 1 to be spinning at maximum speed while fan 2 is off and −10 indicates vice versa. All negative values of thrust are due to fan 2 alone and all positive values due to fan 1 alone.

Comparing the graphs on the left side of the figure, one can notice that on the top left, the control switches between positive and negative, to keep the angular velocity close to the desired value; while on the bottom left, there is no switching: only fan 2 is used all the time. The reason is that there is extra friction for the Table Sat when it rotates counterclockwise (this friction might be a function of the shape of wobble or something else). Although the linear model used does not account for any friction, still it performs well enough in this case (this supports the fact that feedback control

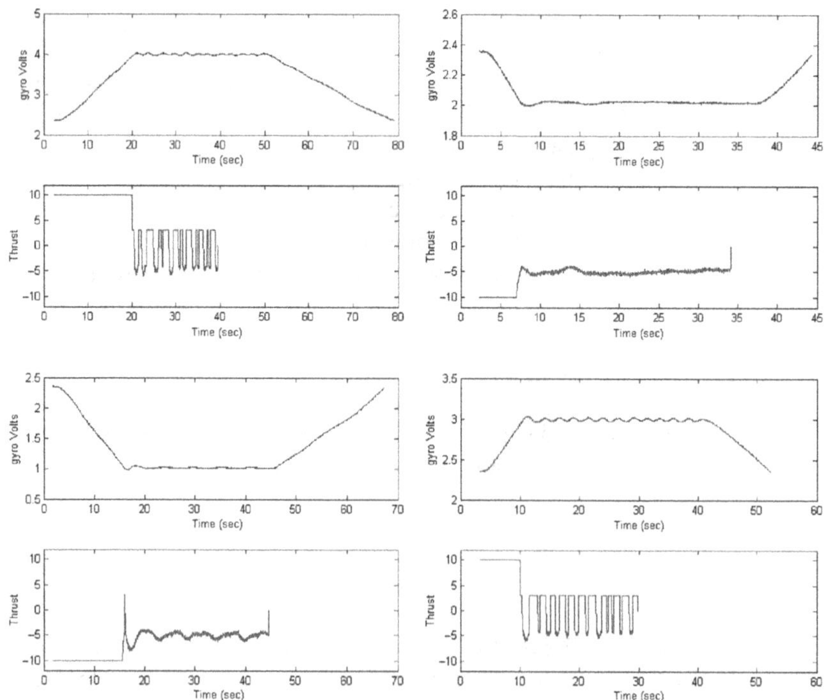

FIGURE 14.2 Velocity control with velocity indicated by gyroscope output voltage.

is robust and immune to small disturbances). Also, one can notice that the extra friction in counterclockwise rotation results in a smoother velocity curve (as opposed to wiggles in clockwise rotation case).

14.4 HEADING CONTROL

For heading control, first, heading is required for which the gyro data can be integrated using Euler's method as follows:

$$\theta_{i+1} = \left[\left(\text{GyroVolts} - \text{Bias}\right) \times \left(t_{i+1} - t_i\right)\right] \times \text{Scaling} + \theta_i \tag{14.4}$$

Notice that the model presented in Equations (14.1) and (14.2) was continuous-time while equation (14.4) is discrete-time. When a continuous-time system was implemented, it had to be discretized using a suitable sampling time (it will be explained how to do this in Kalman Filtering Section). A sampling time of 0.001 seconds was used. The bias in this case was 2.353 and scaling was 48.2385. The resulting heading was in degrees. It was also needed to contain the heading within 0°–360° range so theta was made equal to the remainder of $\theta/360$ whenever it was greater than 360. With Equation (14.4), a heading control was implemented using proportional controller $K = \theta - \theta_{\text{des}}$ with the same threshold technique as in velocity control. The result of heading control experiment is shown below (Figure 14.3).

There are a few things worth mentioning here. First, notice that the initial heading is zero degrees. This is not because Table Sat was at zero degrees when the experiment was started, rather, this was because the initial heading of the Table Sat could not be determined using gyro data, as mentioned earlier. The Table Sat is

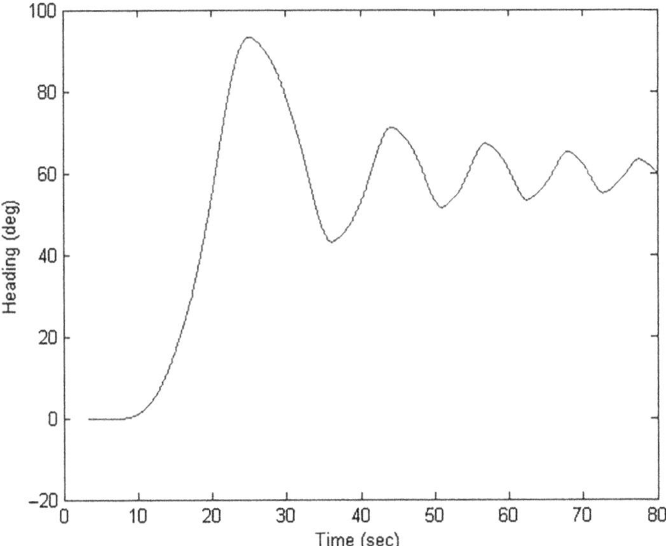

FIGURE 14.3 Heading control using gyroscope.

unobservable using gyro data alone. So, the heading control presented in Figure 2.1 is just a relative heading change of 60° with respect to whatever the initial heading was.

There is another thing worth mentioning here is saturation in gyroscope. The range of gyroscope is roughly 120°/seconds in each direction (counterclockwise and clockwise) while the range of Table Sat is way more than this (varies with friction though).

Before moving on to magnetometers, a final remark about gyro-based heading is that the error in the heading (due to gyro noise) tends to accumulate with time that is evident from the heading data collected when Table Sat was not moving (note that the error could accumulate either positive or negative which introduces additional problems) (Figure 14.4).

Now the calculation of heading using the magnetometer data can be done using three steps:

1. Normalize: $T_{\text{norm}} = T_{\text{gain}}\left(T_{\text{volts}} - T_{\text{bias}}\right)$
2. Angle calculation: $\theta_m = a\tan 2\left(T_{\text{norm},y}, T_{\text{norm},x}\right)$
3. Range adjustment: convert from −180–180 to 0–360 (if theta < 0, theta = theta + 360)

Using the above method, one can get the feedback of heading. This makes the model of Table Sat as single input 2-output. New output equation is as follows:

$$y = \begin{bmatrix} 1 & 0 \\ 0 & 1 \end{bmatrix} \begin{bmatrix} \theta \\ \omega \end{bmatrix} \tag{14.5}$$

Now, using the proportional controller, one can implement the exact heading control (the system is now fully observable with the addition of heading output). Now that heading from two sources is known, it can be seen how well they agree. For comparison of heading from gyro and magnetometer, an initial heading was provided to the gyro data from magnetometer readings.

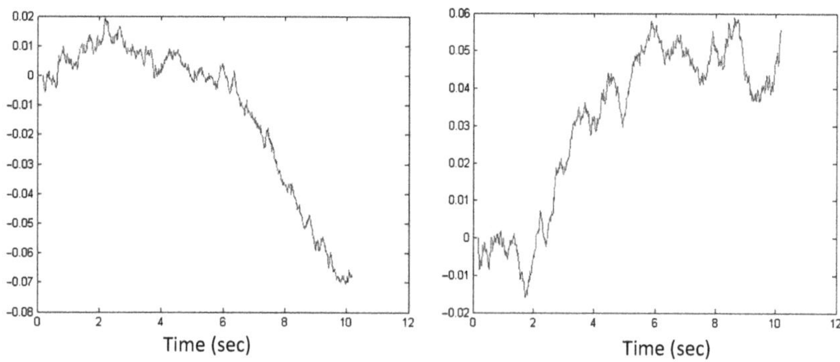

FIGURE 14.4 Gyroscope noise.

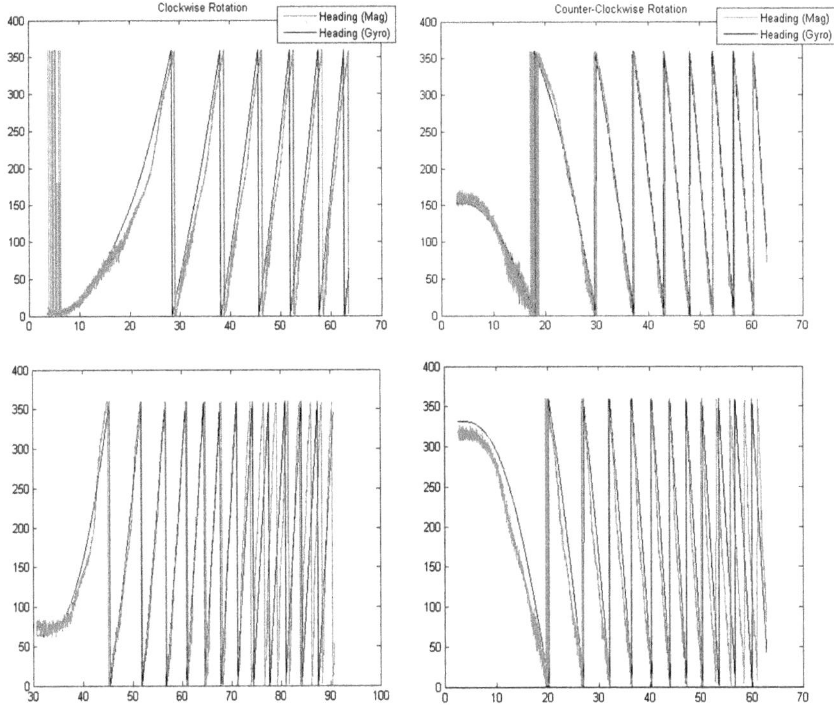

FIGURE 14.5 Comparison of heading with different feedback sensors.

From the top two plots in Figure 14.5 (where the Table Sat spins slowly), the heading calculated from both sources seems to match well. While the plots at the bottom of Figure 14.5 show gyro heading lagging behind the magnetometer heading. This clearly shows the saturation phenomenon in gyroscope (Imagine a one liter measuring cup; if you put any amount of water more than a liter, the cup will show only one liter and the rest will overflow).

Now let us illustrate the more interesting phenomenon, which is the effect of switching the fans on and off on the magnetometer data.

Plots on the top of Figure 14.6 show the magnetometer readings (in volts), while the Table Sat is fixed (cannot rotate) and while plots on the bottom indicate clockwise motion (with and without fans on bottom-right and bottom-left, respectively). In the top-right plot, fan 1 was turned on at $t=5$ seconds, then at $t=15$ seconds, fan 2 was switched on (turning fan 1 off); finally at $t=25$ seconds, both fans were turned off. In the bottom-right plot, only fan 1 was on all the time. It can be clearly noticed that fans introduce disturbance in magnetometer readings. In fact, the percentage errors measured from above plots indicate ex $=0.51\%$, ey $=0.55\%$, ez $=0.467\%$ without fans (ex is error in x-axis, ey is error in y-axis, and ez is error in z-axis measurement), while with fans, ex $=2.58\%$, ey $=1.06\%$, and ez $=2.55\%$. Another interesting thing to notice from these plots is the wobble indicated by z-axis measurements in bottom plots (plotted in red). The most interesting thing here is that apparently fans do not

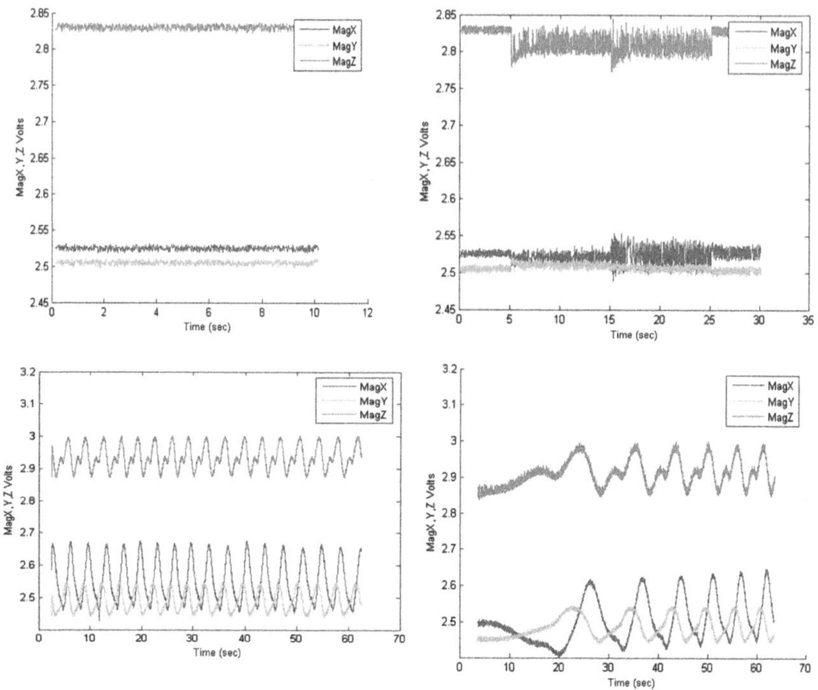

FIGURE 14.6 Effect of fans on magnetometer noise.

affect the wobble. Theoretically, fans should affect the wobble because there should be an exchange of momentum between Table Sat and the fans (possible explanation of this observation is that the mass of fans is very low and almost negligible as compared to the mass of Table Sat and therefore the angular momentum generated by fans is not enough to disturb the motion of Table Sat in a significant way).

Finally, the present heading control plots generated using magnetometer data as a feedback to a proportional controller are presented in Figure 14.7.

One obvious thing to notice here is that the initial condition is the exact initial heading of the Table Sat, and hence this heading control produces the exact desired heading.

Now let us come to calculating heading from the sun sensors. This task turned out to be the most difficult of all since it required consistent light conditions and also it was hard to characterize the behavior of sensors with changing light conditions (while Table Sat rotates).

Theoretically, the following steps were used for calculating the heading from the sun sensors.

1. Normalize $\text{CssNorm} = \dfrac{\text{CssVolts} - \text{CssMin}}{\text{CssMax} - \text{CssMin}}$
2. Convert into degrees (with respect to sun): $\text{CssDeg} = 2\cos^{-1}(\text{CssNorm}) \times \dfrac{180}{\pi}$
3. Compute Heading: $\theta_{\text{Css}} = \text{HeadingCalc}(\text{CssDeg}_1, \text{CssDeg}_2, \text{CssDeg}_3, \text{CssDeg}_4)$

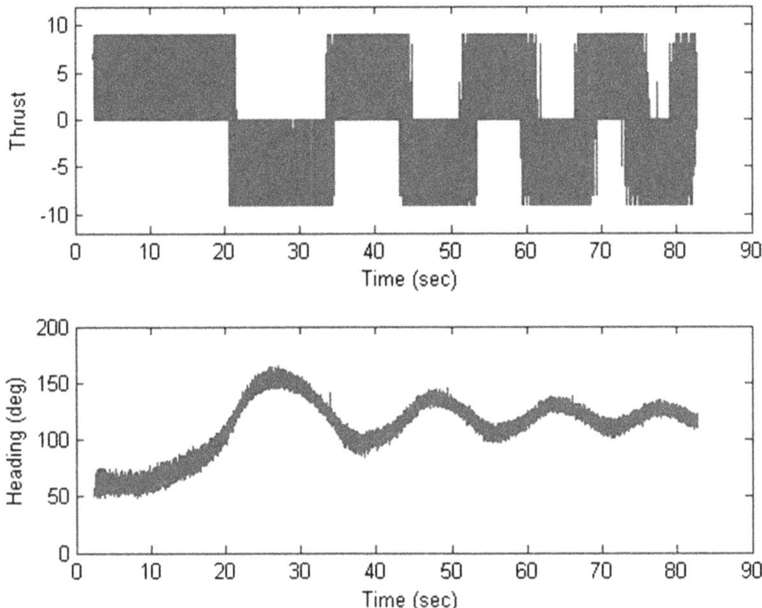

FIGURE 14.7 Heading control with magnetometer feedback.

where CssNorm and CssDeg were calculated for every sun sensor separately and heading calculation algorithm computed the weighted sum of headings calculated by each sun sensor based on which of the four sun sensors was closest to pointing toward the sun (a table lamp in our case). For intuitive sense, the configuration of sun sensors at different positions is shown in Figure 14.8. There are six different configurations based on the criterion of "which sun sensor is closest to the sun." The small arrow in each configuration shows the direction of the sun. Notice that in the range where $0 < \text{heading} < 120$, sun sensor 3 is closest to the sun. The reason why 60° and 120° configuration is shown separately is that during 0°–60° movement of the Table Sat, sun sensor 4 is farthest away from the sun while during 60° to 120°, sun sensor 2 is farthest away. So, the categorization is based on which sensor is farthest and which sensor is closest to the sun. The comparison of headings calculated from magnetometer and sun sensors is shown in Figure 14.9.

There is an interesting feature of plots in Figure 14.9. The plots on top in Figure 14.9 show a comparison of headings while corresponding raw sun sensor readings (in degrees with respect to the sun 0–180) are plotted at the bottom to explain the behavior of heading calculated using these raw readings. In the left column of Figure 14.9, everything seems almost alright except for a couple of short regions where sun sensor-based heading shows a change in noise width. This change comes from switching between different modes of calculating the heading. Notice that at about $t = 20$–25 seconds, sensor 2 is closest to the sun, while at $t = 7$–10 seconds, sensor 1 is closest and at all other times, sensor 4 is closest to the sun. About the column of plots at the right side of Figure 14.9, notice that all four sun sensors are saturated in

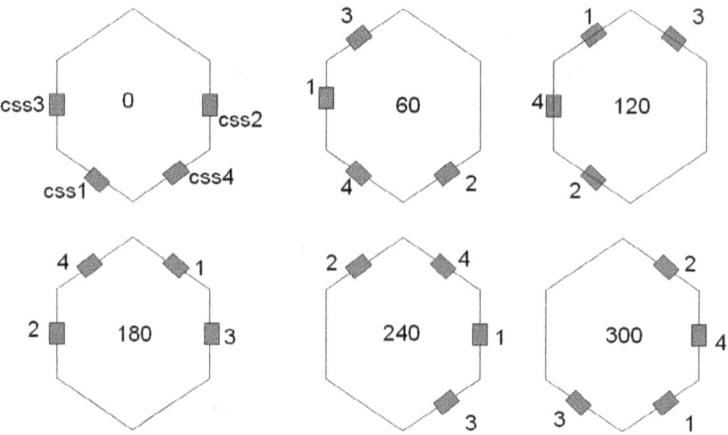

FIGURE 14.8 Configuration of sun sensors.

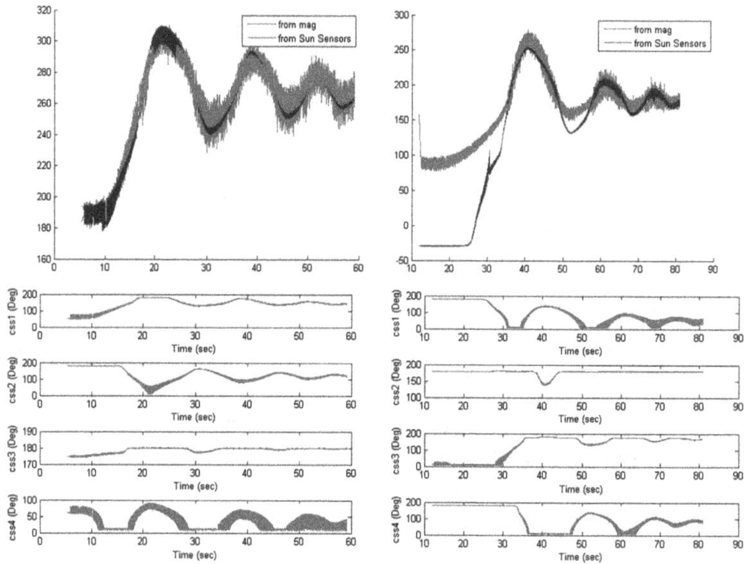

FIGURE 14.9 Heading control comparison between sun sensor feedback and magnetometer feedback.

the beginning which gives rise to poor results in the calculated heading. This shows saturation in sun sensors.

Now that let us move on to estimation using Kalman filter.

14.5 KALMAN FILTER ESTIMATION

Two methods are presented here which are inspired by Refs. [1,11].

14.5.1 Method 1

In this section, the following model of the Table Sat will be used.

$$\begin{bmatrix} \theta_{k+1} \\ \omega_{k+1} \end{bmatrix} = \begin{bmatrix} 1 & 0.001 \\ 0 & 1 \end{bmatrix} \begin{bmatrix} \theta_k \\ \omega_k \end{bmatrix} + \begin{bmatrix} 5\times10^{-5} \\ 0.001 \end{bmatrix} u_{\text{fan},k} + \begin{bmatrix} 5\times10^{-5} \\ 0.001 \end{bmatrix} w_k \tag{14.6}$$

$$y_k = \begin{bmatrix} 1 & 0 \\ 0 & 1 \end{bmatrix} \begin{bmatrix} \theta_k \\ \omega_k \end{bmatrix} + v_k \tag{14.7}$$

where w and v are assumed to be mutually uncorrelated Gaussian random processes with zero means. The covariance matrices of w and v will be represented by W and V, respectively. The above model was obtained from the discretization of the continuous-time model presented previously using zero-order hold method and sampling time of 0.001. The observer model is given by

$$\hat{x}_{k+1} = (A - LC)\hat{x}_k + Bu_k + Ly_k \tag{14.8}$$

where

$$\hat{x} = \begin{bmatrix} \hat{\theta} \\ \hat{\omega} \end{bmatrix}, A = \begin{bmatrix} 1 & 0.001 \\ 0 & 1 \end{bmatrix}, B = \begin{bmatrix} 5\times10^{-5} \\ 0.001 \end{bmatrix}, L = \begin{bmatrix} 0.0133 & 0.0015 \\ 0.0878 & 0.9993 \end{bmatrix} \tag{14.9}$$

Here, L is discrete-time Kalman filter gain calculated by solving the following set of equations:

$$L = PC^T W^{-1} \tag{14.10}$$

$$A^T PA - P - A^T PB\left(B^T PB + V\right)^{-1} B^T PA + Q = 0 \tag{14.11}$$

$$V = \text{diag}\left(3.6\times10^{-7}, 6.4\times10^{-5}\right), W = 10{,}000 \tag{14.12}$$

where P and Q are positive definite matrices. Equation (14.11) is a discrete-time Riccati equation. Using the observer dynamics in (14.8), the heading and velocity of the Table Sat were estimated and Figure 14.10 shows the results. It can be noticed that the estimates for heading are quite precise and represent the maximum likelihood of heading angle given the magnetometer data and assumed noise covariance. These estimates depend upon what sort of noise is assumed and how close it is to the actual noise in the system. Kalman filters do not in general guarantee convergence for nonlinear systems but here, since the linearization is not too far away, the estimates agree with the original data quite well.

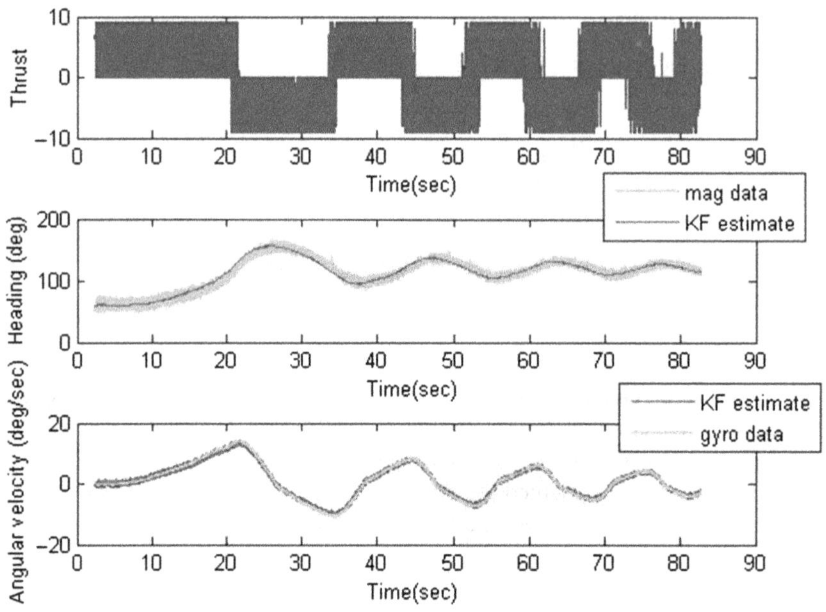

FIGURE 14.10 Kalman filter estimation of heading and velocity with first method.

14.5.2 Method 2

In the above approach, one can notice that Kalman gain is computed only once and put into the system. A slightly more appropriate approach for implementing a Kalman filter uses the recursive calculation of Kalman gain and covariance matrix for states (heading and angular velocity). Once again, Gaussian noise and disturbances were assumed, mutually uncorrelated, and having zero mean values. There are two steps in this method; the first step is the prediction step which uses equations of motion and covariance projection for updating states. Prediction step is given below:

$$\begin{bmatrix} \theta_{k+1|k} \\ \omega_{k+1|k} \end{bmatrix} = A \begin{bmatrix} \theta_{k|k} \\ \omega_{k|k} \end{bmatrix} + Bu_{\text{fan},k} \tag{14.13}$$

where A and B are given by Equation (14.9). The second part of the prediction step is covariance projection, which is carried out by the following equation:

$$P_{k+1|k}^{x} = AP_{k|k}^{x}A^{T} + B_{w}P^{w}B_{w}^{T} \tag{14.14}$$

where

$$B_{w} = \begin{bmatrix} 5\times10^{-5} \\ 0.001 \end{bmatrix}, P^{w} = 0.1 \tag{14.15}$$

Now, the update step which can also be called a correction step based on the measurements at step $k+1$ can be given by

$$y_{k+1} = H\begin{bmatrix} \theta_{k+1} \\ \omega_{k+1} \end{bmatrix}, H = \begin{bmatrix} 1 & 0 \\ 0 & 1 \end{bmatrix} \tag{14.16}$$

The Kalman gain is computed as

$$K_{k+1} = P^x_{k+1|k} H^T \left(H P^x_{k+1|k} H^T + R \right)^{-1}, R = \mathrm{diag}\left(0.06^2, 0.06^2\right) \tag{14.17}$$

Finally, the updated state vector and its covariance are given by

$$\begin{bmatrix} \theta_{k+1|k+1} \\ \omega_{k+1|k+1} \end{bmatrix} = \begin{bmatrix} \theta_{k+1|k} \\ \omega_{k+1|k} \end{bmatrix} + K_{k+1}\left(y_{k+1} - \begin{bmatrix} \theta_{k+1|k} \\ \omega_{k+1|k} \end{bmatrix} \right) \tag{14.18}$$

$$P^x_{k+1|k+1} = \left(I - K_{k+1} H\right) P^x_{k+1|k} \tag{14.19}$$

These two steps were repeated and corrected estimates at every measurement time step were obtained. The results of this method are shown in Figure 14.11.

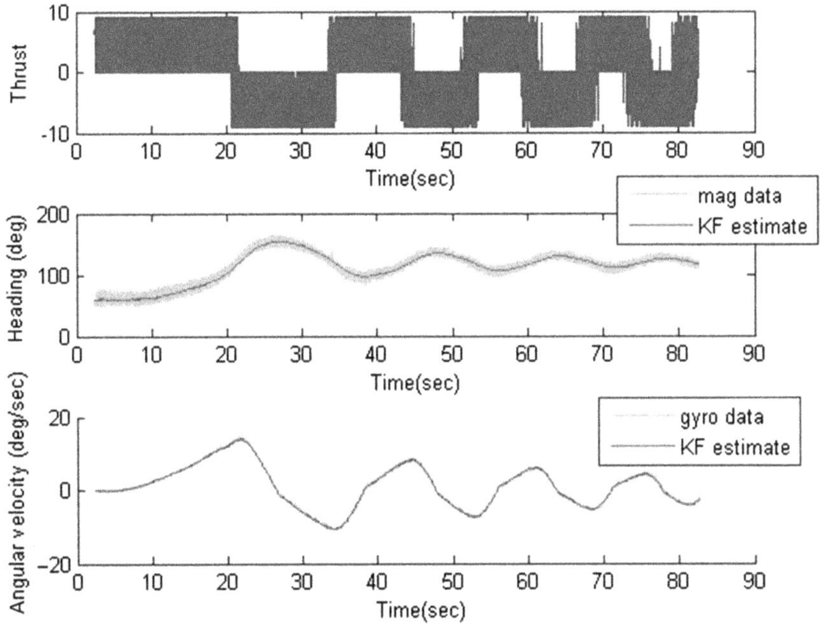

FIGURE 14.11 Kalman filter estimation with second method.

14.6 FAULT DETECTION

A recent study on fault tolerance for spacecraft applications can be found in Ref. [12] but here more focus is on student learning than the research outcomes. Since there are multiple sensors, fault detection can be implemented quite easily (especially for single sensor failures). Here in particular, a method based on triple sensor voting has been used (although there are no three different sensors for position, there are three different ways of computing it if initial conditions are borrowed for gyro-based measurement).

The scheme consists of three layers of errors. Layer 1 is the absolute difference in headings calculated using different sensor data. Error in layer 2 is generated when the error in layer 1 crosses some threshold (there could be many schemes for defining/obtaining the threshold) which in this case is the max of deviations of the calculated headings from the Kalman filter estimate. Error in layer 3 is the triple sensor voting-type algorithm based on errors in layer 2:

$$F_1 = \left(\frac{1 - \dfrac{\theta_{th1} - \left|\theta_{\mathrm{css}} - \theta_m\right|}{\left|\theta_{th1} - \left|\theta_{\mathrm{css}} - \theta_m\right|\right|}}{2} \right) : \left|\theta_{th1} - \left|\theta_{\mathrm{css}} - \theta_m\right|\right| \neq 0 \tag{14.20}$$

$$F_2 = \left(\frac{1 - \dfrac{\theta_{th2} - \left|\theta_{\mathrm{css}} - \theta_g\right|}{\left|\theta_{th2} - \left|\theta_{\mathrm{css}} - \theta_g\right|\right|}}{2} \right) : \left|\theta_{th2} - \left|\theta_{\mathrm{css}} - \theta_g\right|\right| \neq 0 \tag{14.21}$$

$$F_3 = \left(\frac{1 - \dfrac{\theta_{th3} - \left|\theta_g - \theta_m\right|}{\left|\theta_{th3} - \left|\theta_g - \theta_m\right|\right|}}{2} \right) : \left|\theta_{th3} - \left|\theta_{css} - \theta_m\right|\right| \neq 0 \tag{14.22}$$

$$F_{\mathrm{gyro}} = F_2 F_3 \tag{14.23}$$

$$F_{\mathrm{mag}} = F_1 F_3 \tag{14.24}$$

$$F_{\mathrm{css}} = F_2 F_1 \tag{14.25}$$

Here, assumptions include $\left|\theta_{th1} - \left|\theta_{\mathrm{css}} - \theta_m\right|\right| \neq 0$, etc. These assumptions do not create much trouble since the sensor data come from an analog to digital converter and the threshold can be easily defined to a value more precise than the converter used.

To illustrate the working of the above scheme, an experiment was performed where a heading change to the Table Sat was commanded, and during the maneuver, the light was switched on/off for a few times to introduce error in sun sensor-based readings.

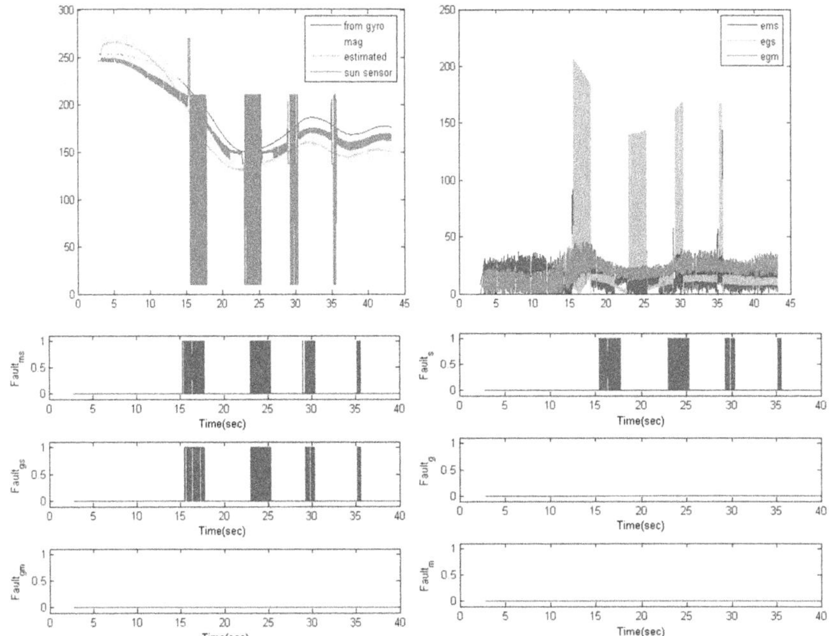

FIGURE 14.12 Results from fault diagnosis.

In Figure 14.12, the top-left plot shows headings calculated during the experiment. On top-right are the layer 1 errors. On bottom-left are the layer 2 errors while on bottom-right are the layer 3 errors which indicate faulty sun sensor readings during the times at which light was turned off.

14.7 CONCLUSIONS AND FUTURE WORK

The work discussed the methods to control the heading angle and angular velocity of the Table Sat using a second-order linearized dynamics model. It also discussed how to obtain heading information out of raw sensor readings. An estimation method based on Kalman filter was presented. Finally, an algorithm based on triple sensor voting that uses 3-layers of errors for the detection of single sensor failure was elaborated.

The chapter indicates how useful Table Sat is for young students in the context of learning theory of feedback control systems and state estimation. Table Sat is not limited to the uses elaborated in this chapter. It can be used to understand various theories (one example of fault detection was presented) related to dynamics and controls.

Slight modifications/additions in the Table Sat can make it even more exciting and interesting. One of those modifications is the addition of reaction wheels attached horizontally with the surface of the Table Sat. These wheels can be used for actuation purposes using the law of momentum exchange. But there are a couple of concerns with the implementation of this idea. First, the wobble in the Table Sat would cause the angular momentum of the wheels to be not exactly vertical, and hence through the exchange of momentum, there could be additional wobble introduced in the Table Sat. Another issue is the possibility of increased vibrations. But these problems by themselves may lead to additional learning.

Strategic and nonstrategic machine tools can benefit enormously with the work presented in this chapter as these commonly utilize a feedback control system.

REFERENCES

1. Vess, Melissa, System Modeling and Controller Design for a Single Degree of Freedom Spacecraft Simulator: *Master of Science*, 2005.
2. Atkins, E., Green, J., Yi, J., Woo, H., Browne, J., Mok, A. and Xie, F., "The Tablesat Platform and its Verifiable Control Software," *In Proceedings of the Infotech@ Aerospace Conference*, AIAA, 2009. Seattle, Washington, US.
3. Bradley, J.M., Clark, M., Atkins, E.M. and Shin, K.G., "Mission-Aware Cyber-Physical Optimization on a Tabletop Satellite," *In Proceedings of the Infotech@ Aerospace Conference*, AIAA, 2013. Boston, Massachusetts, US.
4. Amin, Engr Arslan Ahmed, and Khalid Mahmood Hasan. "A review of fault-tolerant control systems: advancements and applications." *Measurement*, 2019, vol. 143, pp 58–68.
5. Zhang, Youmin, and Jin Jiang. "Bibliographical review on reconfigurable fault-tolerant control systems." *Annual reviews in control*, 2008, vol. 32.2, pp. 229–252.
6. A. Nasir, E. M. Atkins, and I. Kolmanovsky, "Robust Science-Optimal Spacecraft Control for Circular Orbit Missions," *In IEEE Transactions on Systems, Man, and Cybernetics: Systems*, March 2020, vol. 50(3), pp. 923–934, doi: 10.1109/TSMC.2017.2767077.
7. A. Nasir and Ella Atkins. "Fault tolerance for spacecraft attitude management." *In AIAA Guidance, Navigation, and Control Conference*, 2010. Toronto, Ontario, Canada.
8. A. Nasir, Ella M. Atkins, and Ilya Kolmanovsky. "Science optimal spacecraft attitude maneuvering while accounting for failure mode." *IFAC Proceedings*, 2011, vol. 44.1, pp. 812–817.
9. Kaplan, Marshall H. "*Modern Spacecraft Dynamics and Control.*" New York, John Wiley and Sons, Inc., 1976. p. 427.
10. Nise, Norman S. Control Systems Engineering, (With CD). John Wiley & Sons, 2007. Hoboken, New Jersey, US.
11. Leffers, E.J. et al "Kalman Filtering for Spacecraft Attitude Estimation" In *AIAA 20th Aerospace Science Meeting*, 1982, Orlando, Florida.
12. Yin, Shen, et al. "A review on recent development of spacecraft attitude fault-tolerant control system." *In IEEE Transactions on Industrial Electronics 63.5*, 2016, pp. 3311–3320.

Index

Note: **Bold** page numbers refer to tables and *italic* page numbers refer to figures.

For Product Safety Concerns and Information please contact our EU
representative GPSR@taylorandfrancis.com
Taylor & Francis Verlag GmbH, Kaufingerstraße 24, 80331 München, Germany

www.ingramcontent.com/pod-product-compliance
Lightning Source LLC
Chambersburg PA
CBHW060337220326
41598CB00023B/2739
9781032021560